Patrick Moore's Practical Astron

Other titles in this series

Telescopes and Techniques (2nd Edn.)
Chris Kitchin

The Art and Science of CCD Astronomy
David Ratledge (Ed.)

The Observer's Year
Patrick Moore

Seeing Stars
Chris Kitchin and Robert W. Forrest

Photo-guide to the Constellations
Chris Kitchin

The Sun in Eclipse
Michael Maunder and Patrick Moore

Software and Data for Practical Astronomers
David Ratledge

Amateur Telescope Making
Stephen F. Tonkin (Ed.)

Observing Meteors, Comets, Supernovae and other Transient Phenomena
Neil Bone

Astronomical Equipment for Amateurs
Martin Mobberley

Transit: When Planets Cross the Sun
Michael Maunder and Patrick Moore

Practical Astrophotography
Jeffrey R. Charles

Observing the Moon
Peter T. Wlasuk

Deep-Sky Observing
Steven R. Coe

AstroFAQs
Stephen Tonkin

The Deep-Sky Observer's Year
Grant Privett and Paul Parsons

Field Guide to the Deep Sky Objects
Mike Inglis

Choosing and Using a Schmidt-Cassegrain Telescope
Rod Mollise

Astronomy with Small Telescopes
Stephen F. Tonkin (Ed.)

Solar Observing Techniques
Chris Kitchin

Observing the Planets
Peter T. Wlasuk

Light Pollution
Bob Mizon

Using the Meade ETX
Mike Weasner

Practical Amateur Spectroscopy
Stephen F. Tonkin (Ed.)

More Small Astronomical Observatories
Patrick Moore (Ed.)

Observer's Guide to Stellar Evolution
Mike Inglis

How to Observe the Sun Safely
Lee Macdonald

The Practical Astronomer's Deep-Sky Companion
Jess K. Gilmour

Observing Comets
Nick James and Gerald North

Observing Variable Stars
Gerry A. Good

Visual Astronomy in the Suburbs
Antony Cooke

Astronomy of the Milky Way: The Observer's Guide to the Northern and Southern Milky Way (2 volumes)
Mike Inglis

The NexStar User's Guide
Michael W. Swanson

Observing Binary and Double Stars
Bob Argyle (Ed.)

Navigating the Night Sky
Guilherme de Almeida

The New Amateur Astronomer
Martin Mobberley

Care of Astronomical Telescopes and Accessories

A Manual for the Astronomical Observer and Amateur Telescope Maker

M. Barlow Pepin

With 82 Figures

Springer

All illustrations in this book are by the author, unless otherwise noted.

British Library Cataloguing in Publication Data
Pepin, M. Barlow
Care of astronomical telescopes and accessories : a manual for the astronomical observer and amateur telescope maker. (Patrick Moore's practical astronomy series)
1. Telescopes–Maintenance and repair
I. Title
522.2′0288
ISBN 185233715X

Library of Congress Cataloging-in-Publication Data
Pepin, M. Barlow.
Care of astronomical telescopes and accessories : a manual for the astronomical observer and amateur telescope maker / M. Barlow Pepin.
p. cm. — (Patrick Moore's practical astronomy series, ISSN 1617-7185)
Includes bibliographical references and index.
ISBN 1-85233-715-X (pbk. : alk. paper)
1. Telescopes–Amateur's manuals. I. Title. II. Series.
QB88.P46 2004
522.2–dc22 2004049148

Patrick Moore's Practical Astronomy Series ISSN 1617-7185
ISBN 1-85233-715-X Springer-Verlag London Berlin Heidelberg
Springer Science+Business Media
springeronline.com

Printed in the United States of America

Typeset by EXPO Holdings, Malaysia
58/3830-543210 Printed on acid-free paper SPIN 10901120

To my Sun and Stars: Mary, Danielle and Jeannette

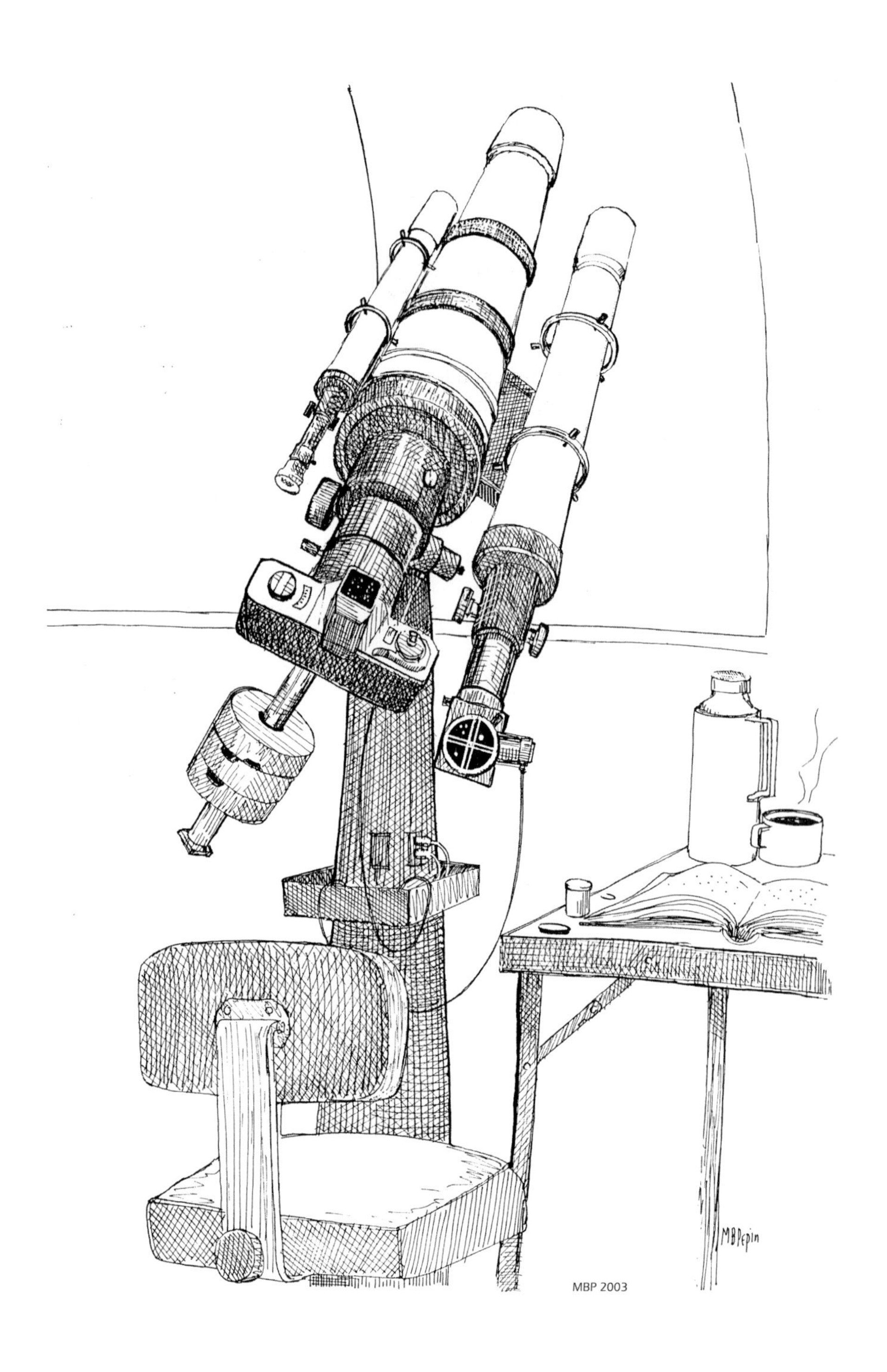

MBP 2003

Foreword

Many new books about astronomical telescopes are published each year, but this one, by M. Barlow Pepin, is unlike any other – at least as far as I know.

The emphasis has shifted lately. Amateur telescope making is less popular than it used to be because professionally made telescopes are much cheaper than they were only a year or two ago. But there is no point in having a good telescope unless it is well cared for, and many observers pay too little attention to this. It is not only a question of common sense; expert guidance is also needed, and in this book M. Barlow Pepin gives it. He goes into detail with regard to all aspects of telescope care and maintenance, and even experienced observers will learn a great deal from it.

Astronomy is one of the very few sciences in which amateurs can make valuable contributions, but good, well-maintained equipment is essential. If you want to get the best out of your telescopes and other astronomical instruments, then this is the book for you.

Sir Patrick Moore CBE, FRS

Preface and Acknowledgments

This book fills a gap left by earlier works of general practical advice, most of which are either out of date or long out of print. It is both a contribution and a tribute to its predecessors' tradition of handing down craft knowledge. Although the volume distills years of practical experience working with astronomical equipment and accessories, it doesn't pretend to be complete in every detail. It would have been illogical to try to set down everything about maintaining equipment for every branch of the pursuit. Such a work would fill an already groaning bookshelf. Furthermore, I don't cover things I have no personal experience with – that's what compendium works are for, and there are some good ones around that cover most aspects of amateur astronomy. Several are included in the Bibliography. Finally, experts continuously cover new developments in fast-changing fields like electronic imaging, optical fabrication methods, narrowband imaging and robotic systems. This work concentrates on the practical acquisition and maintenance of the optics and mechanics that support such endeavors.

Despite the fabulous armchair journeys we can now make on the Internet, using optics under the sky is still what makes the pursuit of amateur astronomy enjoyable. Even the best astronomical equipment requires both care and appreciation to yield good results. Acquiring and caring for it takes patience and a few special skills. There is still time, place and a need for hands-on work.

Years ago I carried a pencil sketch and part of a homemade mount into a machine shop and asked if they could make a simple modification. The foreman in blackened coveralls took a look and said, "Sure, come back tomorrow, it should cost you about twenty." The last time I tried this, a fellow in a golfing shirt looked up from his computer and said, "Sure, send me a CAD file for our CNC guys, and we'll work up a quote for next week. How many units do you need?" Just the software would have cost more than a new mount!

The artisans aren't gone, but they are increasingly hard to find. When you do, they are busier than the devil. Thus, all the more need for a work that covers a little history, with some basic fixes and workarounds for the practical amateur. That's what Section II does. A few processes I have personally worked out and tested are outlined in Chapter 13.

My hat is off as I write this, aware of the many individuals and groups whose collective knowledge is far greater than my own. It was fascinating to review the works of earlier authors around the world, many of whom lived at a time when a telescope was a rare commodity, and most equipment cobbled together in the garages and basements of dedicated enthusiasts. The mission here is the same as theirs: to impart some basics of contemporary instrument acquisition and maintenance to new astronomers, and to provide a useful lookup resource for the old hands. There is always room for improvement. In fact, I fully expect to hear about matters that weren't addressed.

I wish to express sincere appreciation to my present editor John Watson. I am especially gratified for the opportunity here to acknowledge mentors such as series editor Sir Patrick Moore, the late solar observer Donald Trombino, FRAS and editors in the past, including Richard Berry who accepted my first efforts at publication, William J. Cook, Martin Neumann and Simon Mitton. For hints and help over the years I am indebted to Alan MacRobert, Roger Sinnott, supportive members of management and staff at Sky Publishing, and to the Astronomical League, the British Astronomical Society, the Royal Astronomical Society of Canada and the Vereinigung der Sternfreunde (VdS) of Germany. A wide range of amateur and professional optical historians, from Peter Abrahams to Drs. Albert van Helden and Wayne Orchiston, pointed the way to valuable references on historical techniques. Additionally, *kudos* to all the fine people on the business side whose hearts are in the right place, especially Al Nagler who gave open permission to use equipment of his design to illustrate a few basic instruments and accessories for the book. Thanks also to New Focus, Inc. of San Jose, California, and Aerotech, Inc. of Pittsburgh, Pennsylvania, for the loan of bench components for generating laser visualizations of light through the telescope, published before only as line drawings.

Sincere regards to a host of observing companions too numerous to mention, and the many unseen folks, friendly voices in the dark from around the world. Here, too, is a nod to all amateurs' patient spouses and offspring who have lain awake at night wondering "what could possibly be so interesting out there?"

Finally, this volume is dedicated to fellow ATMs and Telescope Nuts, those people who insist on continuing to make contraptions out of odd glass and spare parts that rival or exceed the performance of the best manufactured equipment. The special community of amateur astronomers has endured for centuries. Our activities will doubtless continue long into the future as one of the most fascinating pursuits in this or any other universe. If this volume eases the path for a few, or just helps to pass time on a cloudy night, it will have repaid the effort.

M. Barlow Pepin
Duncanville, Texas
July, 2004

The author. Photo: L. Lattin.

Contents

Section I: **Optical Equipment**

Section I

Optical Equipment

CHAPTER ONE

Treasure in the Cellar

The local astronomical society has formed a committee to inventory its stored equipment. It is a thankless task, one often neglected over the years. The group is spending this springtime Saturday morning down in a damp schoolhouse basement. As they go through the contents of musty cardboard boxes, the society Secretary keystrokes the description of each item into a laptop computer. Some old but useful books are put aside for the next meeting, along with a trunk of telescope making supplies and some usable eyepieces.

With all the boxed items accounted for, the group takes a break. Sipping hot coffee brought by the committee head, they gaze around at the cluttered storage space. There are dusty tripods with broken parts dangling, a mount made of plumbing fittings, a collection of orange electrical cords hanging on pegs. A tall canvas bundle leans against the far wall, covered with cobwebs.

"What's that?" the youngest committee member asks.

"I don't know, let's take a look." The speaker sets down his cup and walks over, unties a few of the brittle wrapping cords and pulls the ragged green tarp aside. Tarnished brass gleams faintly in the dim overhead light.

"Looks like an old 6-inch, but I've never seen it used," says the young member with sudden interest.

"It's sure not on the loan-equipment list," comments the Secretary, swiping at the brass tube with a handkerchief. She sneezes as a cloud of grit and dust rises in the still air.

"Ah, yes, THAT old monster. Its too big to loan. Been down here forever."

The senior committee member looks into space, recalling the details. It seems that a local amateur had bequeathed the refracting telescope to the society sometime after World War II. He remembers hearing that there were problems with it. Nobody's eyepieces would fit in the focuser, and the ones that came with it had narrow fields of view. The mount was heavy and it had an obsolete drive system with weights, chains and whatnot. Furthermore, the tube was too long to fit in any member's vehicle. It was all put "down cellar, and good riddance" when the society acquired a large, modern reflector. Out of curiosity, the Secretary takes her handkerchief to the focusing tube's grimy faceplate. It comes away black. Engraved script shows faintly in the cleaned spot.

"'Cambridgeport'," she murmurs, looking up. "Isn't that near Boston?"

The next society newsletter trumpets a call for volunteers to restore the group's newly rediscovered Alvan Clark & Sons 6-inch equatorial refractor. The following month, we read that someone has located a retired craftsperson with the skill to repair antique brass fittings. Other members schedule a work party to refurbish the frozen drive mechanism. Someone with the skills to safely disassemble and clean rare antique optics is finally located. Then, at the get-together given for First Light, there is general astonishment. Indeed, the old workhorse – newly named for its once-forgotten donor – has cleaned up quite nicely. More to the point, it is a night of good seeing and the uncoated optics resolve details on Jupiter's cloud bands that no one there has ever seen in *any* eyepiece, wide-field or not.

Be it an antique or a solid new production, a few decades is not long in the useful working life of a good telescope. A few "stitches in time" would have saved the society a lot of difficult, if rewarding, work. With proper respect for the design limitations of a 19th-century instrument, the Clark could have afforded many uninterrupted years of fine planetary viewing.

The upkeep of astronomical instruments is, first, a matter of taking basic precautions. Proper storage is only one facet of care and maintenance. Considerations of careful handling and cleaning are equally important in keeping fine optics in working condition. Most optical instruments – telescopes, spotting scopes and binoculars in particular – spend a lot of time outdoors. Obviously, everyone acknowledges the need to take practical steps to control the observable effects of weather on equipment.

Less obvious, but just as potentially damaging, are the insults from the ambient environment. Even a few months of neglect can seriously affect performance and

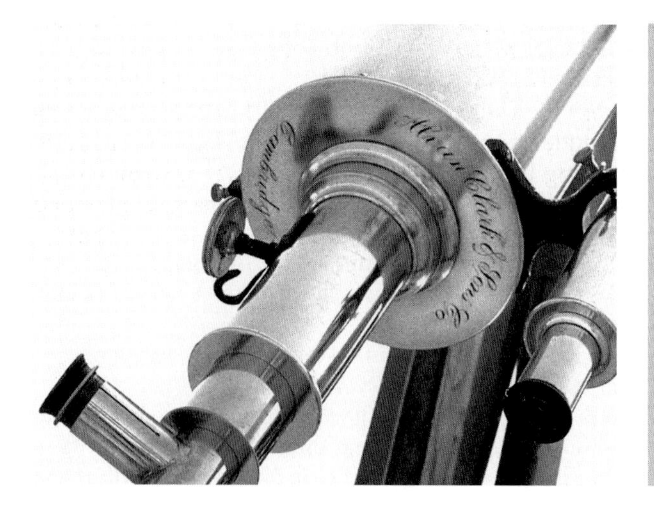

Figure 1.1. Detail of a 6-inch f/15 Clark brass refractor, restored in 1992 by owner Jon Slaton with Steve Sands and Bob Kirschenmann.

usability, depending upon the observing location and the type of use. The results of these damaging factors can harshly affect trade or resale value, important issues for any user with limited equipment funds. Surely we'd rather not dwell on monetary matters where the pursuit of science and knowledge are concerned. Yet, the fact remains that every observer's focus of interest is constantly evolving. New interests require changes in equipment. One of the most efficient means to obtain items that suit one's needs is to trade or sell what you've acquired.

Yet, condition means everything in the valuation of used optics and accessories. This is true even in the case of rare instruments, where one might think integrity and performance would be the overarching concerns. For newer items, "Like New, in the Box" (LNIB) is a common description in the optical resale market. Buyers with today's attitudes might have been called "condition cranks" some years ago, when optics were harder to find, the design selection limited and individual items and parts subject to more variation in manufacture. The old phrase has lost most of its meaning with advances in design and technology. New processes have enabled mass production to ever more precise specifications, and with cosmetic near-perfection. Consequently, buyers in the resale market have increasingly higher expectations as well.

How does this progress affect the way dedicated observers should handle their equipment? Surface treatments provide a telling example. Ironically, the newer synthetic materials and applications produce surfaces that wear more easily than those produced by earlier processes. To give one small example, high-speed screening, printing and coating methods have largely superseded the old engraving-and-filling process for placing makers' names, design types and focal lengths on instruments and eyepiece barrels. Inconsiderate handling wears these newer markings off rather quickly by comparison.

Multicoating for both lenses and mirrors embodies a huge benefit to instrument performance. The drawback is that these are both thicker and softer than traditional simple coatings of magnesium fluoride. With the increased investment represented by even a modest outfit – a selection of oculars for example – careful use, cleaning and storage are more important than ever in preserving both light-throughput and value.

Applied surface armoring made of plastic or synthetic are now generally replacing the leatherette or textured paint surfaces found on higher grades of handheld optics, including astronomical binoculars, monoculars and spotting scopes. These look "sharper" than older designs at first, but unless carefully handled they readily accumulate unsightly wear or staining. There is usually no economical means of replacing the armoring on these newer instruments; a return for factory refurbishing is required, and only worth the charges in the case of a classic. In any case, careless treatment of such products shows immediately, and they will become increasingly difficult to sell or trade in the future.

CHAPTER TWO

Buying It Right: Consumer Strategies

"The Customer is Always Right."

Gordon Selfridge

Twenty years ago, I ordered my first large telescope after telephoning a dozen dealers and manufacturers. One company president set aside a good twenty minutes personally describing his products and business philosophy. I purchased from one of their dealerships that had been run by a man and his son for over a decade. He offered a written "handshake" agreement on the unit, an 8-inch Schmidt–Cassegrain, and shipped it when he received the first payment. When it arrived, I used up another hour or two on the phone getting advice on fine points. He seemed glad to supply it. The man, now departed, loved the pursuit and his place in it.

Nowadays, with profit margins shrinking and competition stiffer by the day, almost no business has that kind of time or confidence in the buyer. That said; the world of amateur astronomy is still a relatively small one. While the industry has shifted into the cyber-age and the Internet has become a nexus for information gathering, buyers still finalize most significant sales of new equipment by telephone. Selfridge's dictum still applies; the way buyers are treated retains a high profile because, in truth, most dealers stock equipment made by the same assortment of manufacturers.

The watchword is "supply," since you can't sell what you don't have. Well-capitalized outlets get their orders in first. They have new stock available and get the lion's share of business. If you know what you are buying, and want to be first to acquire it, such dealers are a good source. Generally, expect quick explanations quoted from factory literature and payment by credit card in advance.

Some buyers prefer to wait until they've see the item in action, had a look through the eyepiece, read the magazine reviews and caught all the scuttlebutt on the Internet before taking the plunge. There's a catch here; manufacturers calculate their first runs of high-tech goodies to meet immediate demand. There can be a wait of several months or longer for the next batch. This is equally true for telescopes and accessories, but particularly for well-marketed electronic and digital imaging innovations.

Moreover, few manufacturers keep identical models in production beyond a year or so. By the time one has found out everything there is to know about an item, it may no longer be available in that iteration. This trend to quick-change models also makes aftermarket service difficult or impossible to obtain, especially when container-loads of equipment from the Far East are often sold by distributors with no in-country technical staff or replacement parts.

The serious observer soon learns to value self-reliance to some degree. Amateur astronomy has always had a large do-it-yourself contingent; hands-on users would rather implement their own solutions than wait weeks for a minor fix. With this need in mind, Part II covers maintenance, and includes a few projects some may find quite useful.

CHAPTER THREE

New Instruments

Tool or Toy?

Most people with a slight interest in astronomy are content spending a few hours a year watching a highly publicized celestial event like a meteor shower or lunar eclipse. The assumption is that the reader of this book already has a growing interest, enjoys observing the night sky or the Sun and intends to put some time into the pursuit. Certainly, it doesn't matter at first whether one can tell Rigel from Betelgeuse (it's the bright reddish star on Orion's *right* shoulder, since the constellation faces us). Simply get a star chart, some kind of optics and, to paraphrase the planetarian Jack Horkheimer, *start looking up!* The time-honored introduction is to attend a few meetings of a local astronomy society that has public observation sessions. You will meet many enthusiasts, and soon know if you want to seriously pursue the activity.

Before I start in on you in earnest (and I will, there's a lot to cover), here are a few hints for those who are fixated on finding a "perfect" telescope right out of the gate. First, seek a range of advice, and spend a little time looking through telescopes before you buy. Another suggestion – regardless of budget – is not to invest too much, too soon. Take a few steps back for a breather before rushing to acquire the latest thing. Someone out there owns a *very* expensive refractor complete with computer pointing, CCD camera and all the "bling-bling" accessories. Problem is they haven't used it since a snap-decision purchase before a highly publicized star party a few years ago. Asked if they didn't want to get the 6-inch out for a look at Mars at opposition, the response was a great big yawn and a fumble toward the remote control for the big-screen TV. Further conversation revealed that learning how to use this ultra-high-tech machinery was too much to handle in a busy work and family schedule.

The lesson is this; keep it simple at first. If you don't have the time or patience for technical wizardry under the stars, it's better to find out now, rather than *after* you've spent a caboodle on razzle-dazzle equipment that will end up as a disused toy.

At the other end of the spectrum, getting shoddy or inadequate goods is another sure way to kill off budding interest. One frequently encounters attractively packaged astronomical telescopes during holiday marketing seasons. The worst of these tend to be refractors of 2- to 3-inch (50–80 mm) aperture, also short Newtonian reflectors in the 3- to 5-inch (80–125 mm) range. They feature shaky mountings, poorly made lenses or mirrors and an assortment of eye-catching but unusable accessories. The eyepieces are usually substandard, with inappropriate focal lengths. Their finding telescopes may have single-element front lenses that give a rainbow image. Some models have computer-pointing capabilities. When the user finally completes setup, motors will precisely slew the telescope to points in the sky where one won't be able to distinguish a planetary nebula from a globular cluster, due to poor optics. The guidelines in this section will assist in distinguishing such travesties from astronomical instruments that live up to the name.

For the novice buyer, confusion starts with ads containing lists of features so complex they border on the absurd – especially for someone who hasn't the faintest idea about things like "1 arcsecond pointing resolution" or "96.8% light transmission at 560 nanometers." There are many helpers at hand: this book and its series companions provide ready reference. Moreover, most astronomy magazines publish a yearly special issue, describing currently available instruments in detail. A browse through a recent issue revealed a comparison chart of nearly a hundred models, a good place to start in the comparison game.

As hinted above, there is such a thing as being *too* concerned about obtaining perfection. One result is a sort of Black Hole of Indecision. Some prospective buyers have simply let years of devoted consumer awareness carry them to extremes of caution. They are constantly seeking the perfect telescope, when they might have spent the weeks of endless looking and "tire-kicking" enjoying the sky in the company of other observers. Still, there are two things to be aware of; the first is a mixed proposition, and the second a pure negative.

Internet groups and lists: The Internet Myth is alive and well in amateur astronomy. Both well-grounded and unfounded opinion circulates with equality in "cyberspace." People who have acquired a rare "lemon" telescope, even disgruntled competitors, sometimes spread misinformation about perfectly good products. Carefully weigh information, check out technical details and get cross-confirmation on strong opinions before making equipment decisions.

Unreliable Dealers: Companies that sell junk for high prices don't stay around long, so buy from sources that have a good long-term record or from newer companies that come well recommended. Equipment rundowns in the periodicals are a good source of information. Find out whether any astronomical journals have reviewed a product, or highlighted it as a new and promising development. Although fairness issues tend to soften some critical aspects, the editors of legitimate journals tend to be experienced observers who take their work seriously.

Responsible editors also vet and drop advertisers if they receive multiple and verifiable consumer complaints. Appendix D contains a short-list of periodicals worldwide that can generally be relied on to supply balanced, well-informed equipment information.

What Dr. Henry Paul noted years ago in his classic *Telescopes for Stargazing* still holds generally true; price-paid is a fair gauge of quality in the optics market. Another good rule of thumb is to buy the best, meaning the best *you* can afford. Buy correctly and from the right source, and you'll even be able to get your money back if you are dissatisfied.[1]

Basic Optical Definitions

Light Gathering and Performance

> To what end then is all the Pains and Trouble in forming and managing Teleſcopes of 30. 40. 50. 100. 200. 300. &c. Feet; When Objects may be Magnified as much by ſmaller Object-Glaſ es, or Object-Glaſ es of ſhorter Focal lengths, combined with Proportional Eye-Glaſ es?
>
> William Molyneux, *Dioptrica Nova*, 1692

Every science student learns the optical relationships that govern the telescope; only experience makes common sense of them. The mind grasps the obvious effects of the telescope with such tenacity that scientific explanations seem counterintuitive. The strongest perception is that the telescope "magically" makes an object bigger and brings it closer to us. Bigger is better. Once this is experienced, the inevitable urge is to increase the effect. Intelligence grants no natural exceptions. After all, the first thing Galileo did in 1609 was to increase the magnification of the Dutch spyglass, impressing the Venetian *intelligentsia* and securing a stipend.

Further development of the instrument revealed the underlying truth; that the diameter of a spyglass's front lens places limits on its magnification. Magnification reduces to the fact that the telescope eyepiece enhances the angle under which we view an object, increasing its apparent *angular diameter*, while maintaining its natural proportions or *aspect ratio*. Thus, the image takes up a larger area of the eye's total field of view, while maintaining (more or less) its true appearance. The earliest observers weren't quite sure whether to think of magnification as an increase in *area* or in *angular size*. They eventually settled on

[1] This book is a practical overview on care and maintenance. Except where noted, examples of equipment are depicted only in order to represent the form factor of the type of item being discussed in relation to general care and maintenance. Observers that are interested in comparing products on a basis of merit or cost are encouraged to consult any number of other publications that give a comprehensive treatment of models according to manufacturer. The special yearly issues of astronomical journals often also include comparative selections and price guidelines (see Appendix D).

angular size as a standard, using the multiplication sign "×" to denote the increase in apparent angular diameter.

The second perception, of "closeness," has made *distance* into another crux of popular misunderstanding. As remarked in *The Emergence of the Telescope*, "How far can I see in that?" is a common question from otherwise technically sophisticated individuals. Intuition aside, we don't see farther in a telescope, rather we perceive dimmer objects, and see them more clearly at *any* given distance.[2]

As Galileo's contemporaries Thomas Harriott and Johannes Kepler realized about the same time, the effect depends entirely on the fact that the front lens of a spyglass is effectively a larger "eye," and that this larger eye sees more. In current terms, the telescope gathers more light energy in the form of *photons* than the eye, making them available to form a brighter, more detailed image on any detector. The larger the lens or mirror, the larger, brighter and more distinct the image can be made, all in geometric proportion to the increase in the size of the telescope's aperture over that of the observer's natural eye. This energy gathering is the telescope's true function: increasing brightness and *resolution*.[3] The second optical part of the visual telescope, the *eyepiece*, is simply a means to this end, and one long abandoned by research astronomers. The effect that so intrigued the human mind and led to the telescope's greater development is now a mere side issue in the professional detection and analysis of energy at all wavelengths.

In visual astronomy, the aperture also determines whether the eye can detect a dim celestial object at all. If the telescope aperture is large enough, sufficient photons from an object pass through the optical system, resolving a perceptible image at the focal plane. If the photon flow is not rich enough, the eye is unable to discern the object within the field of view. We can make a rough comparison with radio reception. Background noise covers the signal from a far-distant station, just as a very dim object remains invisible in a telescope. Increasing magnification, like turning up the radio, makes the noise louder without significantly changing the signal-to-noise ratio. A more sensitive receiving system (larger telescope) is required to hear the station.

In borderline cases, however, you might hear a dim voice punching through the static at very high volume. By analogy, higher magnification *can* increase the likelihood of the eye detecting an object, since pumping up the magnification increases contrast by darkening the background. Looking toward the edge of the field, using *averted vision*, can also amplify the signal to the brain. This technique brings the cone cells of the retina, which are more sensitive to contrast than the central rod cells, into play.

Both effects are dependent on the transparency of the atmosphere, observer visual acuity, the quality of the optical system, and the wavelengths it optimizes. In fact, the ideal telescope would gather information from the entire universal continuum – from gravity, to light, to time – imaging them all in a multidimensional format. This, however, is some years down the road. We concentrate here on observations made in the visual wavelengths of light.

---

[2] *The Emergence of the Telescope*, p. 5 – see the Bibliography.

[3] See *Light Gathering Power* in Appendix B.

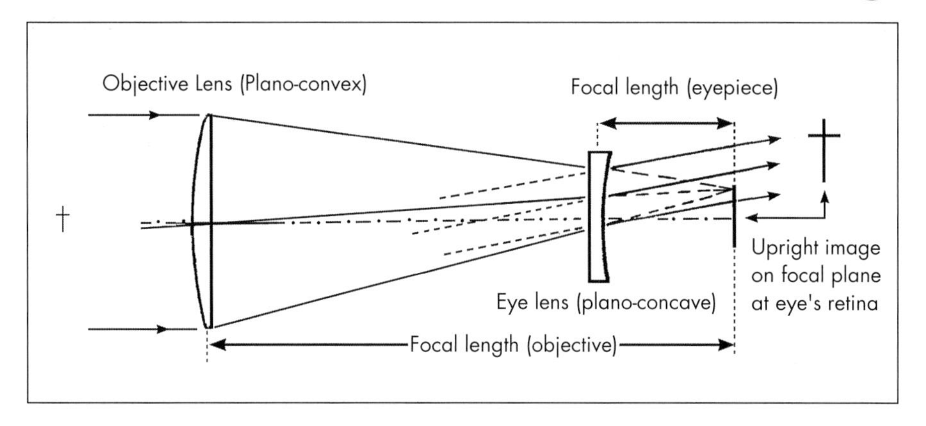

Figure 3.1. The first known telescope was the Dutch Spyglass or "Galilean" Telescope of 1608, with a front convex lens and concave (negative) eyepiece. The design yields images that are upright and correctly oriented.

Aperture and Focal Length

The diameter of the open area of an objective lens or primary mirror that receives incoming light is termed the clear aperture. Most refractor lenses, called objectives, are a fraction larger than this value, because the retaining lip of the cell that hold the lenses covers a few millimeters of the glass perimeter. Some reflectors also have mirrors a bit larger than their clear apertures, due to a constricted front opening or aperture stop.

Every telescope's primary optic focuses light at a specific distance away from its optical center. This *focal length* is usually abbreviated *FL*. With a lens objective, the center point is based on the averaged optical powers of its component lenses. With a mirror, we measure from the central point of the reflecting surface. Focal length is important in practice, since it determines the magnification of a system when using a given eyepiece (or *ocular*, the terms are interchangeable), and the scale of the image produced.

Some telescopes are *compound* designs incorporating additional optics that modify the light as it passes from the primary optic to the point of focus. These types yield an *effective focal length* different from that of the main optic. The Schmidt–Cassegrain system, for instance, uses a convex secondary mirror that *amplifies* the light cone from the main optic, making it steeper and effectively increasing the focal length. Other compound types widen the light cone and effectively *compress* the focal length. The NP101 Apochromat refractor made by Tele Vue is one example. It uses a secondary lens group to compress the light from its objective, yielding a shorter focal length and a wider, flatter field of view.

Focal Ratio

The *focal ratio* is the proportion between an optical system's focal length and its clear aperture (FL:D). To use the eye as an example, the average human has an eyeball with a focal length of about 17 mm. Up through about 40 years of age, the pupil aperture widens to as much as 7 mm in the pitch dark, giving a focal ratio of about 17:7, or 2.4/1 at night. The standard format is f/(FL/D), so one writes "f/2.4." Similarly, a telescope mirror or objective lens with a FL of 800 mm and a clear aperture of 80 mm (800/80 or 10/1) is an "f/10" system. When D is modified, for instance by "stopping down" or reducing the diameter of the primary optic with an aperture stop, observers use the term *effective focal ratio*. A camera lens does this, like the iris of the human eye when reacting to bright light.

The Eyepiece and Magnification

Normal magnification is a basic concept not often encountered in talking about visual observation. It simply means the apparent magnification of the optical system if used without an eyepiece – as in the antique "perspective glasses" that preceded the telescope's invention. We can characterize this value by setting out the relationship between the observer's pupil and the telescope aperture (D/eye pupil).

Figure 3.2. Parallel light passing through a plano-convex objective lens. The focal region where the rays converge can be seen, and the spherical aberration of monochromatic light is evident in the elongated nexus formed by the crossing rays.

A hemispherical plano-convex objective compensates for the high refractive index of the water medium, shortening the system for purposes of practical demonstration. Since the input rays are parallel, coherent light, the afocal layouts in this demonstration series were adjusted to give parallel exit bundles. A 0.5-milliwatt He-Ne laser of 632.8 nm standard wavelength and a beam splitter provided by Aerotech, Inc. were used to project rays for photography in a laser tank of the author's design. Optical element holders used for the project were loaned by New Focus, Inc.

By this standard, a 100 mm aperture and a nominal eye pupil of 5 mm will give a normal magnification of 20×. An observer with sufficient visual acuity can use a telescope without an eyepiece, if in a very limited way. Other than this physiologically dependent value, the primary optic of a telescope has no fixed "power."

One can perceive the natural magnification effect of a telescope by looking into it from the proper distance. Even in small telescopes, for instance, Jupiter appears as a tiny disc rather than a point of light. The unaided eye cannot perform this feat. Strongly farsighted and nearsighted persons may see, respectively, either a correct or an inverted image when doing this. This is because the different "focal lengths" of their eyes cause them to acquire the image from either in front of or behind the instrument's actual focal plane, where it is either inverted or upright. The accompanying laser demonstration photograph shows this "crossing" of the light beams at focus.

Both lens objectives and primary mirrors are technically "objectives," by the original definition, although the term is universally applied to lens optics in practical astronomy. One kind of telescope is a system of optical elements arranged in an afocal configuration. This counter-intuitive term describes the typical setup where the observer focuses the object for viewing. In order to create this situation, we must introduce a second element, as in the "Galilean" telescope or Dutch spyglass, illustrated above. The eyepiece is the essential, second element that creates measurable power, or magnification, which is the increased visual angle an object viewed appears to subtend when viewed through a telescope eyepiece, compared with its appearance to the unaided eye.

The true focal length of an eyepiece is the distance from the optical center of its lens or lens-group to the point at which it focuses. The shorter the focal length of

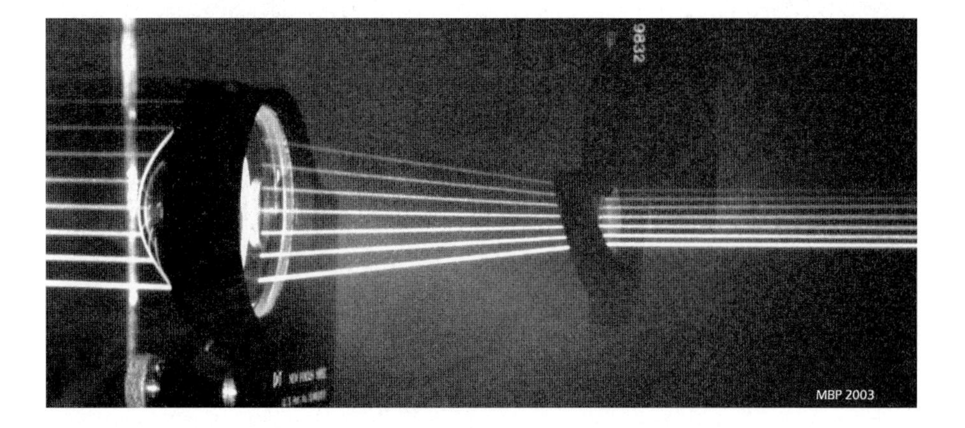

Figure 3.3. The convergence of parallel ray bundles through a simple objective and a plano- concave (negative) eyepiece forms the first telescope, as presented by S. Janssen in Zeeland ca. 1608 and improved by Galileo in 1609. In order to function as a telescope, the negative eyepiece must acquire the light rays ahead of their focal point. The image rays do not cross before reaching the viewer's eye and objects are seen correctly oriented.

Figure 3.4. The convergence of parallel rays through a simple objective to a convex (positive) eyepiece forms the "astronomical" telescope proposed by Johannes Kepler around 1611, probably first used by Jesuit astronomer C. Scheiner for solar projection circa 1630. The image is inverted, as shown here by the crossing of the rays ahead of the eyepiece. Kepler's telescope is longer than the equivalent Galilean system because the convex eyepiece acquires the image at its focal point, beyond the objective's focus.

the eyepiece used in a given telescope, the greater the magnification. Changing the length of the telescope system brings objects at varying distances to focus at a point beyond the eye lens of the eyepiece. Placing the eye at that point conveys the image to the retina. Eyeglasses are unnecessary when the observer has simple near- or far-sightedness, because slightly refocusing the telescope compensates for differences between the focal lengths of the eyeballs of different observers.

In such a system, magnification can be numerically determined by dividing the eyepiece focal length FL_e into the objective focal length FL_0. As an example, the focal length of an objective (for example, FL = 1200 mm) divided by the focal length of an eyepiece (such as, FL = 20 mm) yields the magnification, M. In this case, M = 60 [$FL_0 / FL_e = M$]. Amateur astronomers soon find from experience that almost all telescopes are infinitely variable in power. To paraphrase Lord Acton: with telescopes, power does not necessarily corrupt, but absolute power does corrupt absolutely. The reasons are discussed further on.

Field

Two types of "field" described in connection with telescopic views are the true field and the apparent field. The apparent field of the eyepiece, abbreviated "AF" here, expresses the angle of view subtended by the circle of view within its field stop, independent of the telescope. True field, on the other hand, is the angular measure of the image (the portion of the 360° circle) seen through the entire optical instrument. Binoculars, for instance, typically have true fields between 4

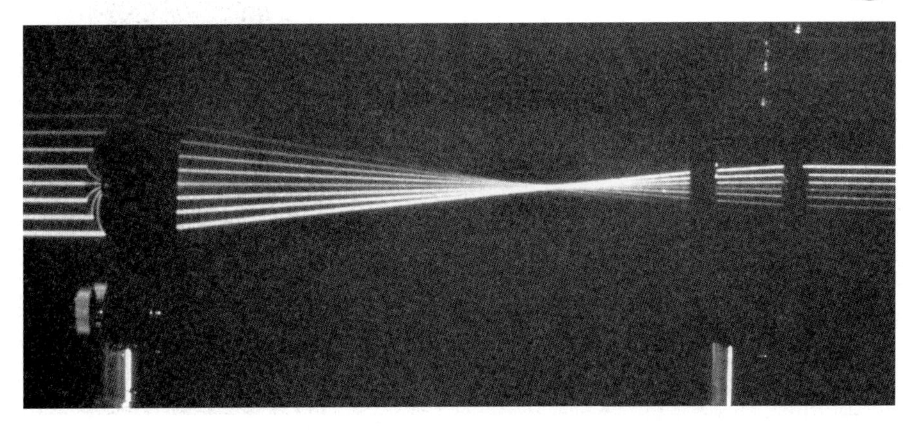

Figure 3.5. This laser tank photo shows the effect of using a Ramsden eyepiece with a Keplerian (astronomical) telescope. This was the second "positive" ocular type developed; the convex surfaces of the field lenses are turned inward (see illustration of Early Eyepiece Types, Figure 5.1).

and 11 degrees. True field is determined in practice by the ratio between the apparent field of the eyepiece used and the magnification it yields in a particular system. A direct visual means of measuring apparent field is described in Appendix B.

Most standard eyepieces have apparent fields between 40 and 65 degrees. So, using the telescope from the previous example, the apparent field with a 22-mm eyepiece (AF = 50°), divided by the magnification of the system (M = 54.54) yields the true field of the system: [AF/ M = TF] which is 0.917°, nearly one degree, or about 1/360th of the circumference of an imaginary Great Circle around the sky.

We find the apparent field angle (size in degrees) of any object observed within the field by multiplying its actual angular size as viewed by the unaided eye by the magnification of the telescope. An object that subtends an actual angle of about 0.5 degree would appear to subtend an angle of 27.27 degrees to the eye when viewed through this system [0.5° × 54.54 = 27.27°]. The Moon is about this size, thus it would take up slightly more than half of the 50° Apparent Field of view in this eyepiece.

To give an everyday example: with the unaided eye, the Moon appears to be about the size of a grape held at arm's length. Viewing it through this system would increase its dimensions to roughly that of a basketball held at the same distance. This is quite an increase in apparent size! Yet, this amount of magnification resides at the low end of the scale normally used for planetary observation.

Another characteristic of afocal systems is that they function in both directions, as when looking through binoculars or telescopes the "wrong way." It may not have occurred to everyone that this silliness reduces the apparent diameter of the object viewed by an exact factor. For example, an objective lens of 12-inch (30.5 cm) focal length paired with an eyepiece of 2-inch (50.8 mm) focal length makes the object appear six times (6×) larger. Reverse viewing reduces the apparent size by exactly this much. This holds true whether the eye lens is positive as in an astronomical telescope, or negative as in a Galilean telescope.

Figure, Focal Length and Performance

The *figure* and *focal length* of its main optical elements are the primary physical descriptors of any telescope. This is true whether it uses lenses, mirrors, or any combination of the two. Figure is shorthand for the "figure of revolution" of an optical surface – i.e. what you would see if a line following the curved surface of a mirror or lens could be extended until it formed a closed loop, then spun about its axis to trace out a solid figure in space.

Properly grinding and polishing a piece of glass with standard methods puts a *spherical figure* on its surface. For the primary mirror of a standard single-mirror reflecting telescope to perform at its best, however, this sphere must be slightly altered by further local polishing of the glass to form a *parabolic* curve. This parabolization – carried out within an accuracy of a few nanometers across the mirror surface – corrects the focused image for the blurriness that would otherwise occur due to *spherical aberration.* The failure of a spherical mirror to create a clear image, as in the diagram here, was theoretically grasped in the early 17th century. Kepler, Descartes, Mersenne, Cassegrain and several others who thought seriously about making a telescope using a primary mirror instead of a lens grappled with the problem. Variations such as paraboloidal and ellipsoidal surfaces looked well on paper, but there were simply no opticians around who could produce them. In those days, optical work had barely progressed to the point of producing accurate spherical surfaces on what were rather small lenses by modern telescopic standards.

Sir Isaac Newton, the namesake of the most familiar type, finally put the concept of the reflecting telescope into practice. Newton's introduction of the *secondary mirror,* an improvement that had eluded his predecessors, made a workable reflecting telescope practical. It has been argued that although he experimented with many figuring techniques, Newton himself never actually achieved a paraboloidal surface in his own optical work in which he did final

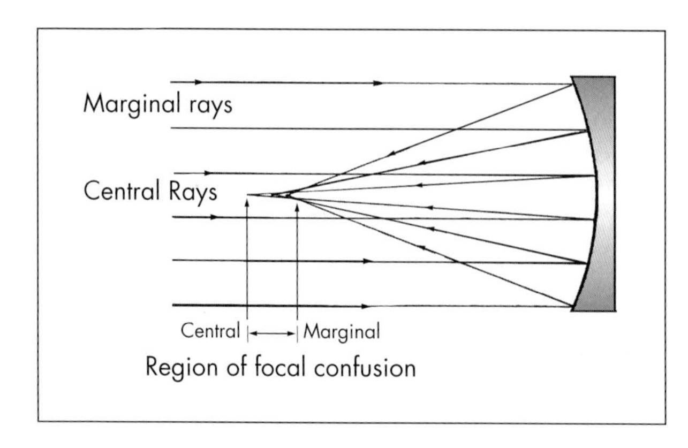

Figure 3.6. Aberration at focus of a spherical mirror. Angle of incidence = Angle of reflection, central and marginal rays focus at different points.

polishing with a paper lap. The paraboloid was probably only realized by James Hadley, some years later. The time and labor required to parabolize a mirror with traditional methods is not trivial. This is why manufacturers still sometimes use spherically figured or "spherical" mirrors for inexpensive Newtonians.

The key to wringing decent performance out of such a setup is to use a mirror with a very long *focal length* in comparison with its diameter. In fact, the *focal ratio* must be at least 10:1 (*f*/10) for a good spherical mirror of around 4-inch (100 mm) diameter. Otherwise, a noticeably blurry image results even at low magnifications. The required ratio also increases geometrically with mirror size. A 15-cm (6 inch) diameter spherical mirror should have a focal length of not less than 200 cm (80 inches) to perform well in visual observation. Consequently, there is no such thing as a good, short standard Newtonian with a *spherical* mirror.

Fast and Slow Telescopes

We call a telescope with a low f/ratio number a "fast" instrument. The term was derived late in the 19th century from the fact that a low f/ratio photographic lens exposes film more quickly. Film speed has increased dramatically over the decades, while average telescope f/ratios have decreased. Therefore, while older sources considered f/15 to f/20 as slow, a telescope is now generally considered slow when the f/ratio is 12 or higher, moderate down to about f/8 or so, and fast when the ratio approaches f/4 or f/5. The user can create a faster instrument from a relatively slow one by adding a telecompressor lens to the optical system. It "fattens" the light cone from the main optic, effectively creating a wider field of view.

The ratio between the focal length of the optical system and that of the eyepiece determines the magnification ($FL_o / FL_e = M$). One should be aware that using a fast, low f/ratio objective complicates reaching the highest useful magnification, requiring the use of very short focal length eyepieces, telenegative optics known as Barlow lenses, or other amplifying systems like the newer Tele Vue Powermate® series.

Practical Limits of Magnification

As mentioned above, a telescope's sole function in the popular mind is to make things look bigger, and the bigger the better. The amateur astronomy literature has devoted much space to countering this impression. As a result, the notion that magnification per se is unimportant – or even that low magnification is always better – has taken root among amateur observers. As G. Dean Williams wrote in his society's Journal, "Like many observers, I associated high powers with dark, shaky, terribly fuzzy images which are impossible to track. A quick glimpse at a deep sky object at 222× from time to time would convince me that sure enough, my short focal length oculars were good for nothing but a little planetary or double star work." As Williams further related in his short article, however, this is an equally gross misconception, and a wide range of powers can be profitably used on most objects.[4]

However, the existence of an upper limit to useful magnification is one of the first things an observer learns from experience. This is especially true if his or her first instrument is one of those 60-mm "refractors from hell" that include a 4-mm eyepiece and a plastic 3× Barlow lens as standard accessories. Under perfect conditions, the aperture of the objective and the quality of the optics define the useful magnification while viewing a typical object containing a mix of light wavelengths in its image. Visual acuity also plays a huge part, since the ability to detect an image at low light levels differs widely between observers. Magnifying an image beyond the threshold dictated by this combination of factors, called *empty magnification*, is simply enlarging a blur.

Aperture and the 50× Rule

The most commonly stated Rule of Thumb for telescopes in general holds that the highest useful power can be calculated by multiplying the aperture in inches by about 50. This works out to about twice the aperture in millimeters. Under very good sky conditions, however, refractor users often find they can profitably use magnifications of 60 to 80 times the aperture (about 3× the aperture in millimeters), even higher for double-star and other specialized work. Higher magnification factors than this are rarely useful. For any particular user, depending upon eyesight, discrimination can often be made at the maximum the glass will bear. I routinely used a 3.8-mm ocular giving 240× in a 100-mm refractor during several Mars oppositions. The glass was able to detect large-scale yellow dust clouds, limb hazes and small variations in the South Polar Cap, while following and sketching their changes. Uncertainties about the scientific interest of such high-power observations with a moderate refracting telescope were allayed when a new yellow dust storm in the Hellas region I detected in this manner was confirmed on the following evening by an experienced patroller using a 16-inch reflector.

Optical theory brings the wavelength of the light viewed into calculating these matters. This is a critical factor for the visual observer employing optical filters. For instance, resolving details of objects may be difficult when using high-blockage nebular or light pollution rejection (LPR) filters. Such filters have relatively low light transmission in the optimal visual yellow-green range. Because the eye is not prepared to discriminate finest detail under these conditions, the telescope may not resolve to its usual limit. Lower magnification than usual may be useful in obtaining a satisfactory view under such circumstances.

---

[4] "In Defense of Power," *Journal* of the Astronomical Society of the Atlantic (Atlanta, Georgia) June 1994.

Exit Pupil and the 5× Rule

The size of the exit pupil of the system, sometimes called the Ramsden disc or "eye ring," is the measure of the smallest diameter of the pencil of light emitted through the eyepiece at focus. The exit pupil decreases in size when shorter eyepieces (yielding higher magnification) are used. You can see the exit pupil by looking at the eye lens of the telescope eyepiece from a short distance when any bright, extended source – such as the daytime sky, the Moon, or a lighted wall – fills the field of view. One can measure it by holding a slip of thin paper, the ground glass end of a microscope slide, or a magnifier with a measuring scale up to the eyepiece and noting the diameter of the bright circle at its smallest point.[5] The ratio between this measurement and the instrument's aperture yields the exact magnification of the system (D/EP = M). The image from the objective is wholly contained within this small pencil of light at the point where the eye observes the field of view.

Calculate the exit pupil of a system by dividing the focal length of the eyepiece by the focal ratio of the system ($EP = FL_e/f$). A second easy method is to divide the aperture in millimeters by the magnification (EP = D/M).

A second observers' axiom, related to exit pupil, is the "5× Rule." The rule generally holds where a) the observer wants to obtain maximum resolution and, b) the full use of the telescope's light-gathering power. In such cases, the minimum useful magnification is about 5× the aperture in inches (D/5 in millimeters).

The eye pupil diameter of a young adult under dark-adapted conditions may expand to about 7 mm. For older observers, or under less than dark skies, the eye's iris contracts, reducing the pupil to 5 mm or less. The maximum exit pupil, by standard calculations, is then larger than the iris of the eye, and effectively blocks illumination from the corresponding outer circumference of the aperture. A corresponding reduction in the telescope exit pupil to this size or less allows the eye to receive the full photon stream from the telescope aperture – i.e. the most brightly lit image. That doesn't mean that one can't appreciate the beautiful wide-field views available at low magnification. In fact, the eye can be kept more easily centered on the field, which is of assistance when fatigue begins to set in. The scope, however, will not reach its limiting magnitude or its maximum resolution, since it is functioning with an effectively "stopped-down," aperture.[6]

Users of scopes with slow focal ratios, typically f/12 to f/20, have less temptation to cope with here than users of faster systems. The common 32-mm widefield ocular used in an f/10 system yields a 3.2-mm exit pupil, well below the limit. The same ocular in an f/4.5 system gives an exit pupil of 7.1, right at the maximum. Since use of the entire aperture is critical in many observations, an unaware observer may waste aperture, for instance, by using the popular 40-mm and 55-mm Plössl oculars with systems of f/5 and faster. As the veteran comet-hunter William Bradfield of South Australia has noted: "...many begin-

---

5 **Please do not use the SUN; the burning rays may cause eye damage, even at a distance.** The exit pupil generated by sunlight well justifies the old term for focus: "burning point."

6 See *Light Gathering Power* in Appendix B.

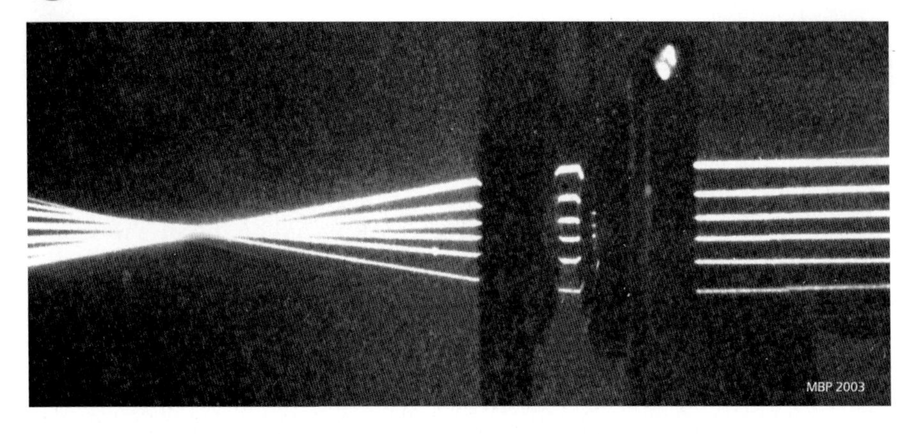

Figure 3.7. Laser tank close-up photo of the pencil of light rays passing through a Plössl eyepiece in an afocal telescope system. The diameter of the parallel array to the right is the size of the exit pupil of the system.

ners in the quest for large fields overlook the various optical relationships and rob themselves of light from the full aperture."[7]

Visual Limiting Magnitude

The concept of limiting magnitude is based on the modern logarithmic scale of stellar magnitudes. This was founded on the ancient Greek notion of star magnitudes first set out by the Greek astronomer Hipparchus in the mid-2nd century B.C. and formalized by Ptolemy three centuries later. The ancient brightness scale ranged from bright First Magnitude stars to the dimmest perceptible luminaries, which he assigned to the Sixth Magnitude.

Investigations by W. Herschel (and later by Pogson) determined that the five magnitude steps coincidentally embrace an increase of light flux or total luminance of about 100 times. Each step could therefore be conveniently rendered on an exponential scale, with 2.512 as the factor. Altair was assigned a magnitude of +1.0, and the general unaided-eye limit was set at +6.5.

Stars are point sources of light, and aperture determines their visibility, down to the visual limiting magnitude of the instrument. Higher powers increase contrast with the background up to a point, and may render dimmer stars visible, as well as other objects imaged as point sources.

In describing the limiting visual magnitude of telescopes, the area of the eye's pupil was originally taken as a standard, and total light flux received through larger apertures related to this in terms of the magnitude scale.

---

[7] Personal communication, 1994.

The attempt (or lack thereof) to exactly determine the amount of light loss through absorption and reflection by the optics complicates the determination of visual limiting magnitude. There is certainly no consensus among instrument makers, who factor in various adjustments when rating their instruments. Factors include slight differences in the assumed size of the eye pupil. The unaided-eye limit used may also range from +6.0 to +6.5. Thus, the limiting visual magnitude of three current models of 4-inch (100 mm) optical tubes are stated as 12.0, 12.5, and 12.7, respectively.

With time, observational evidence has accumulated suggesting that +6.5 is too conservative a figure for the base limit.[8] In well-documented cases, dark sky observations by advanced observers far exceed the listed figures for the apertures used. Thus, the limiting magnitude of a telescope, like the Dawes resolution limit, is a "rule of thumb," and the listing is merely a general guide to relative performance between telescope models of the same line.

Apparent Brightness of Extended Objects

The ratio of the square of the aperture to the square of the focal length determines the apparent brightness of the images of extended objects at focus. Leaving out certain random factors, this can be roughly expressed by the relation $B = (D/FL)^2$. It follows that the brightness of such objects is inversely proportional to f/ratio. By this rule, for instance, a system of 125-mm aperture and focal length 1200 mm has an f/ratio of 9.6 and a nominal "brightness" of 0.011. The same aperture with a FL of 800 mm has a lower (faster) f/ratio of 6.4 and a brightness of 0.024; more than double that of the f/9.6 system.

Given the same eyepiece, a faster objective yields increased brightness, exit pupil and true field, while reducing magnification. Given fast and slow objectives of the same aperture and light transmission, the difference in apparent brightness between images is nullified when eyepieces yielding the same magnification are used. The shorter eyepiece with the fast objective forms the same size exit pupil as the long eyepiece with the slow objective ($EP = D/M$), equalizing image brightness.

Brightness plays a larger part in resolution than the general observer may realize. While aperture alone determines the visibility of point objects such as stars, extended objects suffer increased dimming with magnification. Aperture and optical quality notwithstanding, the eye will not discriminate such details to the telescope's limit in a very dim image. The resolution limit of the eye (about 1–2 minutes of arc, in practice) reduces in this circumstance, through low-light regrouping of the retinal cells into a coarser array. At extremely low light levels, enlarging the details by increasing magnification will not help, since the image will only be further dimmed in the process, leading to "empty magnification."

---

[8] Luginbuhl and Skiff, *Observing Handbook and Catalogue of Deep Sky Objects*, p. 9 (see the Bibliography), state that the visual limiting magnitude from a building on the (well-lit) University of Arizona campus was generally about 6.2, increasing to visual magnitude 7.0 at the nearby dark-sky site routinely used for other observations.

Here again, differences between observers' visual acuity play a large role in what is apprehended.

The Airy Disc

Astronomer Royal George Biddle Airy (1801–92) modeled the visual telescopic image of a star in the course of his investigations of optics. The so-called "Airy disc" or diffraction disc is the small central portion of the false image of a star formed by a telescope at focus. Light not contained in the disc forms neat, concentric diffraction rings, or Fresnel rings surrounding the disc. The size of the star image is proportional to the wavelength of light, and inversely proportional to the aperture of a particular optical system. Thus, the larger the aperture, the smaller the Airy disc in stars of the same color.

This ideal representation has confused some observers, who assume their optics are flawed when star images on the Airy model do not appear during routine sessions. Perhaps the very heavens stilled when the doughty Astronomer Royal approached the eyepiece! Due to atmospheric turbulence and "local seeing" disturbances of the air in and around the telescope, the model appearance is rarely glimpsed in the field. Typically, one sees an amorphous central discoid surrounded by a series of broken, shifting ring segments. It also appears spiked with diffraction rays where linear obstructions like spider vanes are present, as in a Newtonian or Cassegrain reflector.

In years of observing, the author has only experienced a handful of occasions when the atmosphere rendered a perfect Airy model visible – all in the wee hours of still, humid, subtropical mid-summer nights of marginal transparency. At times like this, planetary detail stands out in the eyepiece like lines on a banknote, and doubles generally seen as barely split seem to have widened to admit an extra measure of black space between their components. The Airy model can also be examined under still conditions when viewing a suitable artificial star or properly sized illuminated pinhole through a small telescope at a distance of 20 meters or so.

Viewed through a good, unobstructed objective, the disc by definition contains nearly 85% of the total luminance from the star. The small amount of light scattered from the disc of this diffraction image partially explains the pinpoint stars and high contrast seen in a refractor. The disc itself even grows smaller with increasing obstruction. Most obstructed systems, such as Newtonian reflectors and Schmidt–Cassegrain telescopes, contain anywhere from 75 to as little as 50% of total light in the Airy disc, with the rest of the light thrown out of the disc, into the surrounding rings and background.

Resolution and Dawes' Criterion

In a field where there is someone prepared to defend an ironclad answer to every question, it is refreshing to note the confusion of overlapping definitions for the important factor of *resolution*. Since we put telescopes to so many uses, there is a gulf between criteria. Does one use point sources like stars or extended sources such as planetary surface details? For example, we can define the resolving ability of a telescope for observational purposes as the true *angular size* of the finest detail observable. We must then decide whether to take the human eye, with its maximum resolution of around one arc-*minute* for our gauge, or accept some other criterion such as photographic emulsion grain with its molecular-sized clumps, or CCD pixels sized in microns, yielding discrete data that can be digitally optimized.

The devil is also in defining what constitutes significant detail, since we can also frame performance in terms of separation of close line pairs or *linear resolution*, which is said to be mainly dependent upon focal ratio, or upon observable contrast between adjacent areas, angular size largely notwithstanding. For which instruments and applications is one or the other more relevant? Optical analysts now divide the matter into three functions; CTF (Contrast Transfer Function), MTF (Modulation Transfer Function) and OTF (Optical Transfer Function) in conjunction with computer-generated graphical simulations, in order to compare optical performance on three levels of analysis.[9]

As stargazers, let us look at practical criteria, since there is no "ideal sky," and assume that all things being equal, resolution is dependent primarily upon D, or aperture. Under the *Dawes Criterion*, sometimes referred to as *Dawes' limit*, the resolution theoretically attainable is succinctly defined for practical purposes. Pre-eminent double star observer The Reverend William Rutter Dawes (1799–1868) worked it out based on the separation of "white" double stars of comparable magnitude (~6 visual). Dawes' rule, in practice, defines the resolution attainable with an excellent 6-inch achromatic objective. He arrived at it by testing various excellent telescopes of different aperture, including his 6.5-inch Merz, 7.5- and 8.25-inch Clarks, and an 8-inch Cooke with which he pursued double-star measurement and cataloging from 1831 on.

The simple equation for Dawes' limit is: $a = 11.6 / D$

Where D is the aperture of the objective in centimeters, a is the measure in arc-seconds of the smallest separation between double stars of near-equal brightness that can be positively identified as double by elongation of the image.[10] An excel-

---

[9] Suiter, pp. 209–15 explains the CTF in detail as it relates to extended objects – see the Bibliography.

[10] Measuring in the prime visual wavelength mentioned previously; the yellow-green at 5500 Ångstroms. (One Ångstrom unit (Å) = the standard measure of electromagnetic wavelength, equivalent to 10^{-5} cm.). Wallis and Provin, pp. 80–90 (see the Bibliography) show in modified computer simulation that the visually apparent radius of the ideal diffraction disc on the Airy model increases from 0.6 seconds of arc at 4000 Å to 1.2 seconds at 7000 Å in an unobstructed system of 6-inch (150 mm) aperture. Very red stars cause difficulty in critical visual estimations of brightness. Such bloating phenomena may also affect the acquisition of faint, close companions in strongly colored pairs. In general, however, the relative difference in a due to wavelength factors in the visual range is indistinguishable to the eye. Such considerations have greater relevance in photographic or electronic detector resolution. The apparent size and brightness of stellar images vary widely according to the spectral response of the particular emulsion or detector array, which usually exceed the range and/or the limits of the eye.

lent 5-inch (130 mm) objective, for instance, will theoretically reveal the presence of a double star down to a separation of 0.89 arcsecond. The criterion is among the strictest, a difference in peak intensity of only a few percent near the endpoints of an elongated Airy disc.

Dawes lived up to his characterization by Airy as "the eagle-eyed." We could add fuel to the reflector–refractor debate by noting that Dawes had independently discovered the inner "crepe" or C-ring of Saturn with his 6.5-inch Merz refractor, a standard achromat, whereas William Lassell, possessor of a 24-inch f/10 reflector, was disbelieving of its existence until Dawes patiently pointed it out to him in the Merz.[11] Lassell's telescope, the first large equatorial reflector and a marvel of its age, was at an apparent disadvantage in contrast, whereas its light grasp abetted his discovery of three dim "point sources" – satellite Triton of Uranus (1846), and the two largest moons of Neptune, Ariel and Umbriel (1851).

Note that makers of objectives and especially mirror surfaces are capable, in general, of far better corrections than in Dawes' era. Systems that beat the diffraction limit are consistently fabricated in critical conditions of manufacture. Under cool, semitropical December skies with excellent seeing conditions, the author detected doubles of separation 1.1 arcsecond (32 Orionis – elongated), 1.4 arcsecond (52 Orionis – "hourglass" diffraction pattern) and 1.7 arcsecond (28 Orionis – cleanly split) in a 100-mm refractor, with the stars near upper culmination. Moreover, the first two pairs were downright impossible on the following night, under only slight turbulence.

An excellent 4-inch Apochromat refractor *does* have the ability of resolving 1.1 arcsecond double stars under careful scrutiny. Thus, Dawes' criterion holds for excellent seeing, while many other observations confirm that image degradation by atmospheric turbulence and turbidity due to suspended water vapor or dust particles limit resolution under most circumstances, regardless of telescope quality.

Wavefront Error and the Rayleigh Limit

To touch on physics, the consistent spacing of phase in the light waves that have passed through a telescope system is an important indication of its optical quality. An instrument with significant faults in this regard cannot possibly resolve to its theoretical aperture limit, even under ideal conditions. According to

[11] In a letter to a friend, George Knott, Dawes pointed out that "William Lassell using his 24 inch reflector had closely scrutinised Saturn on the 21st November, a very fine night, finding no suspicion of the dusky ring."

"*... On the December 2 Mr. Lassell came to see me and the next night, the 3rd being fine, I prepared to show him the novelty which I had told him of and explained by my picture; but, naturally enough, he was quite indisposed to believe it could be anything he had not seen in his far more powerful telescope. However, being thus prepared to look for it, and the observatory being darkened to give every advantage on such an object, he was able to make it out in a few minutes.*" Jeff Hall, History of Science and Technology Group, University of Liverpool (Michael Oates, History of Astronomy website)

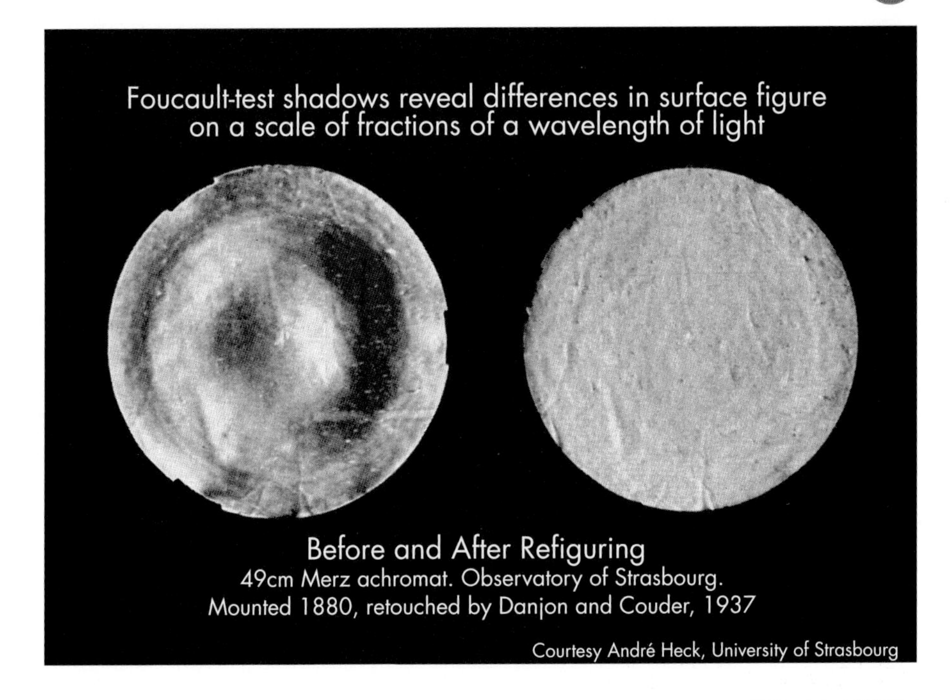

Figure 3.8. Foucault test photograms.

Trinity College physicist, the Lord Rayleigh (J. W. Strutt, 1842–1919), "The wavefront emerging from a good objective is contained completely between two spheres whose radii differ by a maximum of a quarter wavelength." This is a practical limit, since Rayleigh arrived at this 1/4-wave figure from the study of the diffraction patterns of star images in excellent instruments of his time.[12]

Optical systems or elements that meet this "1/4-wave test" are often described as diffraction limited. The objective or mirror set is assumed to have a figure and surface polish such that the errors induced in a light wave passing through the system do not exceed 1/4 of the wavelength of yellow-green light. This is equivalent to 0.0000555 cm or 0.0000219 inch, a tiny quantity indeed! Commercial makers of mirrors sometimes abuse this limit in citing quality, usually applying it only to the surface characteristics of the primary optical element. When such statements come without individual test results, take them with a pinch of salt.

Optics are best tested by a combination of methods, depending on the application for which they were designed. Observers may characterize errors in optical systems at the eyepiece by the use of visual estimates like the star test, or capture the results with film, video, or digital imaging for analysis. Unless one is a

[12] The mid-nineteenth century was a time of rapid progress for visual astronomy and optics in general. In fact, most of the figures mentioned in this section were aware of one another's work, argued conclusions, and collaborated in observations on occasion.

qualified, seasoned expert in star testing or other visual methods, however, analog or digital interferometry is required to closely estimate the wavefront error of a system. A popular analog field test uses the "Ronchi screen," a small plate or film containing a pattern of equidistant parallel-ruled opaque lines introduced by the Italian optician Vasco Ronchi.[13]

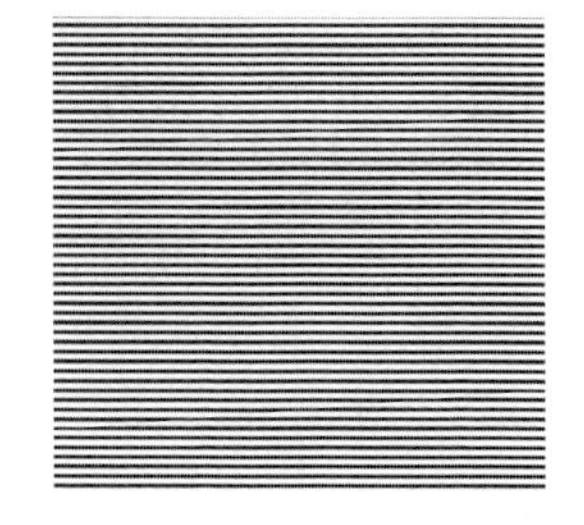

Figure 3.9. 40-line Ronchi screen.

Opticians Armand Fizeau, Léon Foucault and others first developed other such tests in the 19th century. The classic Foucault knife-edge test is still the basic diagnostic tool used during figuring of optical surfaces and for determining their final character. Existing optics are often refigured to improve their performance, which can be compared as in the illustration here, which shows improvement in the figure of the Merz objective of the Great Refractor at Strasbourg after retouching by the expert opticians Danjon and Couder. The photographs were taken under identical Foucault test conditions, with the same apparatus.

Laser interferometry: Computer analysis and laser technology allow much more complex optical testing, involving measurement and plotting of the wavefront errors of reflection or transmission at thousands of discrete points, creating a virtual map of the focal plane at very high resolution. Graphics can represent these results well. The RMS (root mean square) standard applied to multipoint testing averages the errors over the entire surface, diagnostically more revealing than the traditional methods that measure peak-to-valley (P-V) wavefront error by averaging the results from a dozen or fewer concentric zones.

Another diagnostic standard combined with others for comprehensively describing the performance of an optical system is Strehl Intensity, a determination of the proportion of total luminance from a point light source contained within the Airy disc at the focal surface of the system. Optical perfection aims for a Strehl Intensity of 90% or higher. The illustration in Figure 3.10 shows the graphical output from such a test.

System Components

When amateur astronomers use the word "telescope," they are usually referring to the complete assembled instrument, from tripod foot to optical tube. It is helpful to learn the divisions of the mechanical parts in order to avoid confusion (or the embarrassment of resorting to vague descriptions like "that metal thing with the knobs on it").

---

[13] Vasco Ronchi, On the Shadow-stripe Method in the Study of Light Waves, *Zeitschrift für Instrumentenkunde*, v.11, Nov. 1926. Ronchi discusses the characterization of the distortion in the image of a ruled glass reticle with 50 dark lines per millimeter viewed at the focal plane as an elegant method of quantifying aberrations of telescope objectives and mirrors.

Most telescopes have three main components: (1) The optical tube assembly, (2) the mount and (3) the supporting structure. The latter is usually either a tripod or a pier (sometimes also called the pillar or plinth). In portable designs, these parts separate easily for transport or storage, usually by removing a few hand-tightened bolts.

Wave Front Error and Definition

As stated earlier, a system output error of $\frac{1}{4}$ wavelength in peak visible light (yellow-green) has often been considered the sine qua non. While Rayleigh (and some modern authors) have considered this amount of wavefront error as representing the smallest defect noticeable to the eye, it has often been argued that solar system and double-star observers use better optics to great advantage under the best of conditions.

Dawes' limit – being a practical and eyesight-dependent threshold of distinguishing between equally bright adjacent point sources of the same color – is not directly applicable to the observation of extended objects. A truly excellent telescope under the best conditions can exceed both Dawes' limit in stellar observation and Rayleigh's limit in perception of fine contrast detail in extended objects, such as in the differentiation of shadings on the Lunar and Martian surfaces.

Although telescopes can be better now than ever before, the truth is that too few have compiled objective results of observations using truly diffraction-limited and tested optics to make the required number of confirming observations to redefine such rule-of-thumb criteria. We must look to scientific sources for confirmation of what our eyes tell us. Texereau felt that the Rayleigh criterion for optical quality was too lenient when applied to excellent observing conditions.[14] This is not just an expression of the perfectionism one might expect from an optician of his caliber. In his writings, he thoughtfully brings up Danjon's, Couder's and Françon's researches, emphasizing the fact that any wavefront error in an optical system is simply added to the degradation of the image caused by other negative factors such as atmospheric turbulence.

In the experience of Françon, barely perceptible planetary features of (measured) 0.03-wavelength contrast difference were detectable in a 1/16 wavefront-error system, suffering a 30% loss of visibility when viewed in a system with $\frac{1}{4}$-wave error. To make a serviceable extrapolation: Effective degradation by atmospheric turbulence and local seeing that reduces a 1/16-wave system's planetary image to an effective $\frac{1}{2}$-wave (tolerable), would degrade a $\frac{1}{4}$-wave objective's image to approximately $\frac{3}{4}$-wave (noticeably fuzzy). This doesn't apply to routine stargazing, but it does refute the widespread notion that it is useless to seek out 1/16-wave quality optical systems for use under a so-called $\frac{1}{2}$-wave sky.

Optical Tube Assemblies in General

The basic parts of the optical tube include the primary optic in its cell, the focusing assembly and the main tube. The reflector or compound telescope usually has reflective and sometimes lens components mounted along an optical axis, contained in the tube assembly. Provisions for attachment to the mount (mounting rings, yoke, dovetail track or mounting

[14] Jean Texereau, abstracted in Twyman, pp. 577–78 – see the Bibliography.

block) may be part of the assembly or part of the mount, depending upon the design.

Wave Front Error and Definition, cont'd

Thus, while we can appreciate the modern empiricists' position that Dawes' and Rayleigh's work are only handy benchmarks for characterization of system quality, they still have relevance in characterizing achievable telescope performance for the practical observer.

Refractor Tubes *The Objective*: The objective is the lens combination at the front of the refractor tube. It gathers light and creates the image. Most usable refractors have a pair or more of lenses comprising the objective. The basic form using "normal' glasses is called an achromat, but more sophisticated designs are also common.

Objective Cell: The elements of the objective are assembled into a cylindrical machined metal (sometimes plastic or other material) housing or *cell.*

Main Tube: The main telescope tube is a cylinder of lightweight metal, usually soft aluminum. It holds the mirror and carries the focuser.

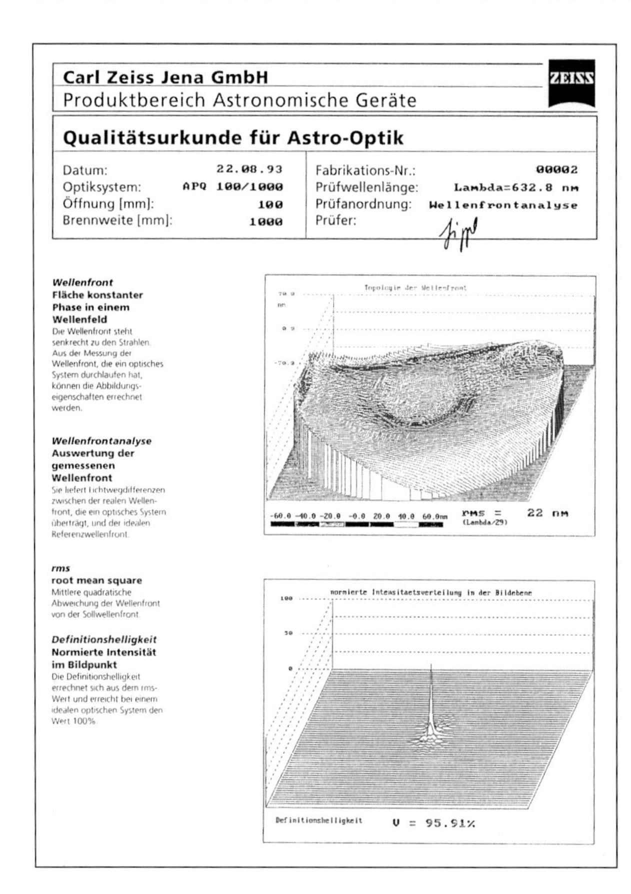

Carl Zeiss Jena GmbH ZEISS
Produktbereich Astronomische Geräte

Qualitätsurkunde für Astro-Optik

Datum:	22.08.93	Fabrikations-Nr.:	00002
Optiksystem:	APQ 100/1000	Prüfwellenlänge:	Lambda=632.8 nm
Öffnung [mm]:	100	Prüfanordnung:	Wellenfrontanalyse
Brennweite [mm]:	1000	Prüfer:	

Wellenfront
Fläche konstanter Phase in einem Wellenfeld
Die Wellenfront steht senkrecht zu den Strahlen. Aus der Messung der Wellenfront, die ein optisches System durchlaufen hat, können die Abbildungseigenschaften errechnet werden.

Wellenfrontanalyse
Auswertung der gemessenen Wellenfront
Sie liefert Lichtwegdifferenzen zwischen der realen Wellenfront, die ein optisches System überträgt, und der idealen Referenzwellenfront.

rms
root mean square
Mittlere quadratische Abweichung der Wellenfront von der Sollwellenfront.

Definitionshelligkeit
Normierte Intensität im Bildpunkt
Die Definitionshelligkeit errechnet sich aus dem rms-Wert und erreicht bei einem idealen optischen System den Wert 100%.

Figure 3.10. Custom makers often provide test data for the optics they supply, something rare in mass-produced instruments. This laser-interferometer test record from the Carl-Zeiss Jena firm records the root-mean-square wavefront error across the full aperture of a 4-inch (100 mm) f/10 APQ objective, deriving a Strehl intensity >95%.

Rear Cell and Focuser: The rear cell (or "tailpiece") fitted into the back end of the tube is generally of cast and machined metal. It holds the focusing tube, or tubes, and thus maintains the proper alignment of eyepieces and accessories with the optical axis.

Reflector Tubes The Primary: The main element is typically a mirror with either a spherical or a paraboloidal figure.

Primary Cell: A cell holds the primary mirror in position in most reflecting designs. It may be as simple as a flat disc of composition board, or as complex as a framework of specially configured alloy or composite material rods linked to a series of pressure pads underneath the mirror surface and joined to a cast or machined framework. The basic function is the same; it holds the primary mirror firmly in place, and determines the angle at which it directs its converging light toward focus. Compound telescopes have the mirror in an arrangement that includes a cell or cylindrical sleeve-mount, usually at the base of the main tube.

Main Tube: The main reflector tube can be a cylinder of lightweight metal, usually soft aluminum. It holds the objective and carries the tailpiece and focuser. In the case of Dobsonian reflectors, the "tube" may actually be a structural truss of rods, usually based on the Serurrier model of stiff triangles, but not always. Sometimes a simple array of stiff, straight tubing locked firmly in place is used.

Secondary and Focuser: In simple reflectors like Newtonians, the secondary flat mirror folds the light 90° to the focal point near the "top" end of the tube, through the focuser tube which holds eyepieces and visual accessories. In compound designs, there is usually a cell in the bottom end of the tube, generally of cast and machined metal. It may also hold the focusing tube, or tubes, in the general position of a refracting tube. Alternately, the mirror itself moves to focus.

Mount, Tripod and Pier (Plinth)

Mount: The mount holds the optical tube steady and facilitates the telescope's motion (slewing) to a chosen object or area. Manufacturers usually employ either the alt-azimuth or the equatorial style of mounting with commercial telescopes.

Tripod: The tripod, a three-legged wooden or metal supporting frame, is the most widely used portable base for all optical instruments which must be held steady. Users of telescopes, cameras and surveying equipment usually mount their instruments on tripods. Although many have tried variations, such as the "Quadra pod," designers have yet to devise a more stable solution for general portable use.

Pier or Plinth: Mounting piers are also available from many manufacturers. Usually fabricated of steel pipe, 3 to 6 inches (80 to 150 mm) in diameter, they hold the instrument at a generally usable height. Adapting one's observing position becomes necessary. Variable-height piers are available from some custom makers. Drilled and/or tapped holes at the top and base allow attaching the mount head and base fittings – or a leg set in the case of a "portable pier."

A Beginner's Telescope

There are only three readily available telescope optical systems generally suitable for the raw beginner. Most others require enough knowledge of optics to support a steep learning curve for acquisition, adjustment and maintenance. These basic designs are the *Newtonian reflector*, the *refractor* and *compound telescopes*. The last category includes the Schmidt–Cassegrain ("SCT"), the Maksutov–Cassegrain ("Mak") and the Maksutov–Newtonian ("Mak-Newt"). We will concentrate on these.

Other types of excellent design, such as the "Classical" Cassegrain, the Ritchey–Chrétien compound, so-called Hyperbolic Astrograph, Schiefspieglers, the Houghton and other exotic concoctions are specialized instruments, largely unattainable for the beginning observer in today's mass market. When available, they are custom-made and therefore hit the pocketbook hard, too.

Why a telescope? Discussions about starting out often raise the question, "Why start with a telescope at all?" Seasoned telescopists advise devoting sufficient time to learning the heavens with the unaided eye and good binoculars before beginning telescopic observation. This is good advice indeed, since observing with binoculars is pure fun, and can be a challenging activity.

Back in 1985 I knew enough about glass to realize the old 7 × 35 Bushnells I had on hand were probably sadly inadequate for the task of spotting an 8th magnitude Comet P/Halley near the Pleiades long before perihelion. Still, I persevered, lying on the tar-sticky road (the only open sky near my house at the time) and brushing away mosquitoes. Acquiring the tiny mothball of Halley's in the sky two nights in a row with that beat-up old pair of binos (and noting that it had changed position), still rates as one of the best feelings I've ever had in astronomy.

Experiencing the big picture is an economical and rewarding way to begin. Nonetheless, few serious observers continue stargazing long without *some* sort of telescope. The question remains: which instrument to acquire when it comes time to observe dimmer objects at higher resolution?

Refractors, covered here first, were the original type of the telescope. They are regaining their former great popularity as instruments for both beginning and advanced amateur astronomers. If visual study of the mysterious objects of the deep sky – galaxies, faint nebulae and dark structures – appeals to you as the most exciting realm for exploration, an economical reflecting telescope is likely the ideal first instrument. If photography and imaging is your aim, catadioptric or "cat" telescopes provide unique advantages beyond economical aperture. The mounting chosen for the optical tube is of great importance, since it must be stable and track smoothly enough to allow long exposures.

Computer-pointed motorized units are increasingly available. Still, one might carefully assess the temptation to leap immediately into the whirlwind of computer-aided observation or – to paraphrase the noted German optician–observer Thomas Baader – to risk viewing more and more objects, but with ever-blinder eyes.

There is no one "best" telescope for all types of observation. In fact, most avid astronomers eventually end up by acquiring *more* than one, using each for the work to which it is best suited. First, let's look at cost, and at the heart of the basic telescope systems: their optics and optical tube assemblies.

A Note on Historical Cost Relationships

"There are no admission fees posted in the sky"

Anonymous Stargazer

Looking back, we can easily demonstrate why stargazing has maintained popularity as an economically feasible pursuit. With apologies to amateur telescope makers everywhere, we will need to sacrifice a "sacred cow": the prized homemade reflector. In 1960 Neal Howard touted the costs savings for a home-built Newtonian, writing, "Your completed telescope probably has cost somewhere between $50 and $350. ... Most amateurs find that they have spent around $125 for their completed instruments ... a professionally made 8-inch reflector, complete with drive and setting circles, costs from $600 upwards. Its approximate equivalent, a 4-inch refractor, starts at $950."[15]

Adjusting these figures to mid-1990s consumer purchasing power in real dollars gives a cost range for the homebuilt telescope of between $240 and $1650, averaging about $600. Looking at the "store bought" prices, we see the financial incentives operating at the time. For an 8-inch equatorial reflector by Cave, the adjusted cost is $2900 (and up), with a staggering $4500 (and up), for a 4-inch equatorial refractor with Fraunhofer achromat by Unitron of Japan.[16] It is no wonder cries of "profligate!" and "snob" were often directed at refractor buffs.

However, let us bring the amateur artisan's viewpoint up to date. Howard's $1650 investment for a top homemade 8-inch will easily obtain a 10-inch commercial reflector today, with *no* labor involved. For what the "professionally made" reflector cost, you can now get a computer-pointed, fork-mounted 10- or 11-inch Schmidt–Cassegrain, with Global Positioning System, an electronic alignment compass, and a database that automatically slews the telescope to tens of thousands of objects with arcminute accuracy or better. Finally, for the same investment as the bulky, achromatic f/15 Fraunhofer, one can easily obtain a fast new equatorial 4-inch Apochromat, with half the tube length and perfect color corrections. No refractor approaching this quality was even *available* in Howard's day.

There's more: The quality of telescopes is slowly, but continually ramping up, while prices drop. A continental optician recently related that his firm had tested some Apochromats imported from the Far East, and found them disturbingly good in terms of price and quality versus the finest mainstream units currently available.

There is a new middle ground, too. Equatorially mounted 5-inch (125 mm) and 6-inch (150 mm) refractors with very decent image quality, 2-inch (50.8 mm) focusers (and *adjustable cells*, something long absent from low-range productions) are now available for about what a decent 3-inch (80 mm) f/11 on a simple altazimuth mount cost only a few years ago. Furthermore, optical correcting

---

[15] *Standard Handbook of Telescope Making*, p. 238 – see the Bibliography.

[16] Data from Warner and Ariail, p. 33, based on Statistical Price Abstracts 1990–93 (Consumer Price Index) – see the Bibliography. Models of the era identified by the author from advertised magazine prices.

Figure 3.11. A 4-inch f/15 Unitron equatorial refractor has a well-corrected achromatic objective of the Fraunhofer type.

filters that cancel a large amount of false image color, available from U.S., German and Japanese sources, render the performance of such units superb in comparison with low-end productions of the past.

These and other factors make a small or mid-size refractor an easily defendable choice for the beginner, regardless of budget. Let us start with the refractor then, as we examine the basic types.

Types of Telescopes

Refracting Telescopes

When it hears the word "telescope," the public usually imagines a long tube that one looks into the end of. This has to do with decades of cartoons depicting wacky astronomers and oversized lenses poking out of domes. Basic astronomy manuals, however, seem especially fond of pointing out refractors' higher price-per-inch versus reflectors as a sort of ultima ratio. Many beginners follow this seemingly unassailable financial logic, and make a small reflector their first instrument.

On the other hand, if we were to base recommendations on say, "pennies per day of longevity," leavened with intangibles such as ease of use and image contrast, the refractor could easily come out ahead. In fact, past surveys have shown that the majority of current serious amateurs, many of whom have gone on to professional work, began with humble 60-millimeter refracting telescopes. Their simple construction makes them durable and intuitive to use. Many of the good old ones are still around, while the majority of their reflecting cousins have succumbed to disuse – if not through coating spoilage or loss of small fittings, then to the frustration and tedium of mirror adjustments attempted by small, untrained hands.

Performance Sidgwick mentions in his classic tome *Observational Astronomy for Amateurs*, "Regarding type of instrument, it is generally agreed that for critical definition of an extended image a refractor is superior to a reflector of equal aperture and optical excellence, owing to the difference in the diffraction patterns of the two instruments."[17] This is nothing new, but readers of earlier manuals will find widely varying estimates of just how, and in what way, refractors outperform reflectors. For example, Henry Paul states in *Telescopes for Skygazing* that "... a 6-in. [15 cm] reflector, at half the cost, will under good conditions perform as well as a 5-in. [12.7 cm] refractor and have a marked advantage for astrophotography."

Paul is by far the most conservative source found, and his remarks must be qualified. Moreover, he disagrees completely with Sidgwick, who felt that the advantages of refractors, unlike that of reflectors, *increased* with aperture, whereas Paul states that "the advantages resting with the refractor diminish rapidly" as one moves from a 3-inch to 8-inch (80 mm to 200 mm) size.[18] Other respected authors, such as Roth, make the gap wider, comparing a 4-inch glass to an 8-inch mirror.

In any case, with new and sophisticated refractors on the market – not to mention electronic imaging – most of the drawbacks that sparked controversy in the past have simply evaporated. Concerning Paul's remarks on photography, a fast, well-corrected Apochromat using modern film will now record the same extended objects in a fraction of the time, and over a wider, flatter field.[19] The "performance gap" between refractors and reflectors has consequently widened.

Comparative field experience over the years shows an equivalence at the high end – between an Apochromatic refractor (see below for a discussion of types of objectives) and a high-quality Newtonian reflector of the same focal ratio

---

[17] p. 76 – see the Bibliography.

[18] p. 65 – see the Bibliography. By good conditions, Paul means primarily steady ground-level temperatures, since thermal effects were the bugbear in an era when Pyrex was rare even for diagonal flats, and most makers of amateur reflectors used standard plate glasses to fabricate mirrors.

[19] With respect to photographs: a film of ISO (ASA) 64 or 100 was considered "fast" in the 1960s, and the best refractors available to amateurs were standard achromats (Fraunhofer or Steinheil designs) with focal ratios of f/12 to f/15 or higher. Thus, film exposures at refractor prime focus took four times (or more) longer than with a typical amateur reflector of f/8 or faster. As to the quality of images taken at the time: those favoring Newtonians for photography generally used only the image at the center of the film frame, since coma-free, wide-field photography was the province of costly instruments – the astrographic doublet, the classical Cassegrain, Maksutov or Schmidt camera.

adjusted to optimal tolerances – to be roughly 4 inches (100 mm) versus 7 inches (180 mm).

While viewing the Moon through a 6-inch Apochromat refractor, the former Association of Lunar and Planetary Observers (ALPO) Jupiter observing chairman once remarked to me, "I am seeing detail here that I can't resolve in my 10-inch [Newtonian]." Observers have long laid full blame for this difference in image quality at the feet of the secondary-mirror obstruction in reflecting and catadioptric systems. In fact, as Canadian experts Dickinson and Dyer determined in a concerted round of shirtsleeve testing, deliberately obstructing the objective of a 7-inch Astro-Physics Apo refractor up to a level of 43% was not sufficient to significantly degrade the image quality, other than to reduce contrast.[20] Testing reveals that the doubling of surface errors by reflection, mirror surface roughness (including "micro-ripple" revealed by phase-contrast testing), thermal density gradients and air currents in the tube, diffraction from spider vanes and other variables all count. A good modern objective does support higher magnification than a good mirror of larger aperture due, as Sidgwick noted, to the *overall* difference in diffraction pattern.

Are there equalizing factors that can reduce the gap? Certainly there are. Given good optics, the single most important is precise mirror *collimation*, just the area where the beginner lacks expertise. By the strictest definition of a non-aberrated field, the mechanical alignment tolerance of an f/6 Newtonian, considered rather "slow" in these days of large, fast mirrors, is a mere 0.017 inch (0.43 mm) of deviation along the optical axis at the image plane. It is much less in a faster system like the typical f/4.5 Dobsonian. Let collimation drift off more than this, and point images start to warp and become asymmetrical, extended images rapidly lose contrast. Chapter 10 covers adjustment routines that will give good results, optimizing reflector observations.

Careless abuse aside, a refractor objective is less prone to incidental harm than most mirror components. First, the thick cell with its dewcap or glare shield provides mechanical protection. A practical consideration for family use is just this fact – refractors are even relatively "kid-proof," since there is less chance of optical damage from things accidentally dropped or spilled down a long optical tube. In addition, the eyepiece is usually nearer to the ground, avoiding tipsy stools or holding children up to a lofty eyepiece. Lens objectives also stand up well to atmospheric conditions and humidity that can destroy the delicate mirror surfaces and fragile secondary mirror mountings of reflectors. There is another advantage for all-round nature study: A small, portable refractor is outstanding for daylight observation of birds and animals in the field.

Achromats Just how does a refractor work? A bit of optical history is in order here. Early opticians using single lenses for 4- to 6-inch (100 to 150 mm) objectives had to use focal lengths up to 100 times the lens diameter (or even greater) to achieve barely acceptable images. Amateur optician Dan Chaffee found in a recent experiment that a single convex-plano crown glass lens of only

[20] "Telescope Myth #8" in *The Backyard Astronomer's Guide* – see the Bibliography.

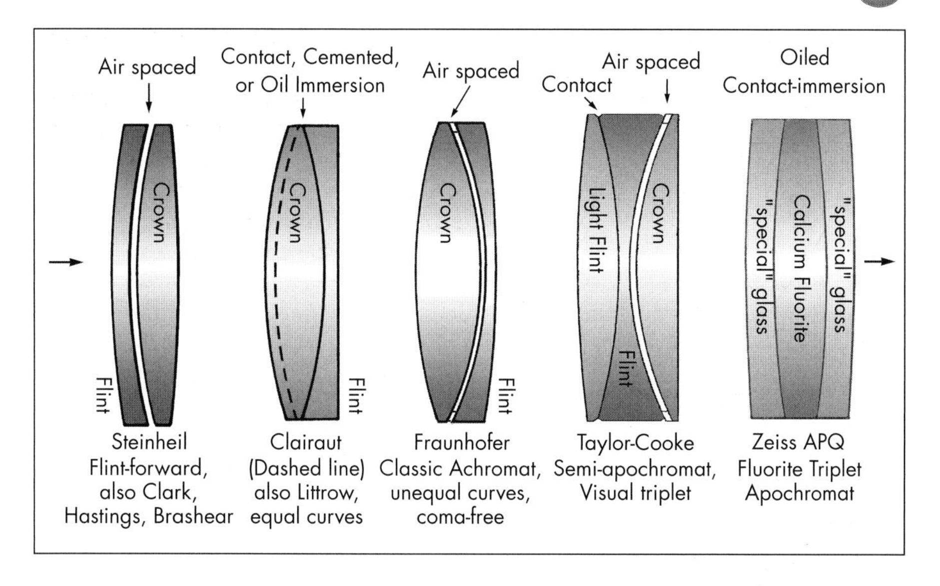

Figure 3.12. Refracting objective profiles; some significant types, curves exaggerated for clarity.

50-mm aperture needed to work at 68 times its focal length in order to render relatively sharp images at 55×.[21] A good achromat, which is the generic term for high-quality refracting objectives of the traditional type, produces excellent images at a focal distance of only 10 to 15 times its aperture.

One constantly encounters the terms "flint glass" and "crown glass" tossed around in descriptions of lenses as though they were more specific than they really are. They are antique terms that once described the two main kinds of household glass.

Old crown glass was simply the clear substance typically used for windows and drinking glasses, made using powdered silica (siliceous sand), soda (sodium or potassium nitrate) and a small amount of metallic oxides, with additions of small amounts of various substances such as arsenic or rock salt as clearing agents. Better grades incorporated potash.

Flint Glass was (and still usually is) denser than crown, embodying a higher proportion of metals, such as lead, often in the form of litharge (red oxide) and manganese, with calcined, powdered flint substituted in various proportions for sand, and potash added. Its higher density gave it different breakage characteristics, and it "caught the light" in a different way. The clearer and more brilliant flint glass found more decorative uses, such as cut crystal and chandeliers. Modern glass types with high dispersion, (i.e. with Abbé dispersion numbers around 50 and below), are considered part of the flint family.

[21] *Amateur Telescope Making Journal*, issue 16.

Simply put, a lens made of regular window glass deviates (refracts) light rays at a certain angle depending on its index of refraction and its surface curvature, bringing them to a nominal focus. It also spreads or disperses the red and blue light in the spectrum by about 1/50th of its focal length. The spectral dispersion through an identically shaped element made of a flint glass, such as lead crystal, is significantly greater.

The flint glass chosen for standard achromats also deviates the light more, due to a higher index of refraction. Thus, ideally, a telescope objective made by combining a positive (convex) lens of crown glass and a negative (concave) lens of flint glass of the right relative optical powers will cancel out each others' color errors, bringing two of the primary colors of an image to focus at nearly the same point. English experimenter Chester Moor Hall was first to determine and exploit this fact, around 1733. London opticians John Dollond and his son and Peter continued Hall's work in the 1750s, with John publishing his work in 1758 and subsequently receiving a patent on his achromatic ("color free") objective.

The term is clearly a misnomer, since strong purple or green hazes around objects were common, but it was a magnificent improvement over the old single lenses. The full solution is more complex than it sounds, and 18th century opticians did not arrive at it. Theoretically, the longitudinal color aberration (with the blue rays focusing ahead of the red ones) would be perfectly corrected if crown and flint glasses dispersed the spectrum in direct proportion. Due however, to what opticians dub the "irrationality of the spectrum" or partial dispersions, the proportion of dispersion in different regions of the color spectrum differs for different types of glass. For most optical glasses, the differential between crown and flint is greater in the blue-green to purple range than in the red. This makes the selection of glass types a matter for endless computation, and perfect color corrections in a refractor something of a Holy Grail until comparatively recently.

Chromatic difference of magnification or lateral color compounds the problem, increasing with the field angle at which light bundles entering the lens combination form the image.[22] The field of even a good achromat will still be somewhat curved, focusing differently at the edges than at the center. Even the best also have some residual astigmatism – the condition where light is focused along two different planes at right angles, say horizontally ahead of focus and vertically beyond focus, leading to stretching of point images out of focus, and a slightly cross-shaped point image when focused on the optical axis. Add to these the concerns of controlling off-axis coma and spherical aberration (fulfilling the aplanatic condition as defined by Zeiss optician, Ernst Abbé), along with other more subtle aberrations of the higher orders. The reader will begin to realize just how much science has gone into the making small discs of glass work as a telescope!

When the correct glasses are chosen, and the thickness, spacing and surface curves of the lenses are all precisely calculated, the different color images in an

[22] This is a drastic effect in single lenses, and the achromat must be corrected for it as well. Otherwise, the portions of the image produced by rays peripheral to the optical axis create an image that is increasingly larger in the red than in the blue at focus, progressively blurring the image in proportion to its distance from the axial center.

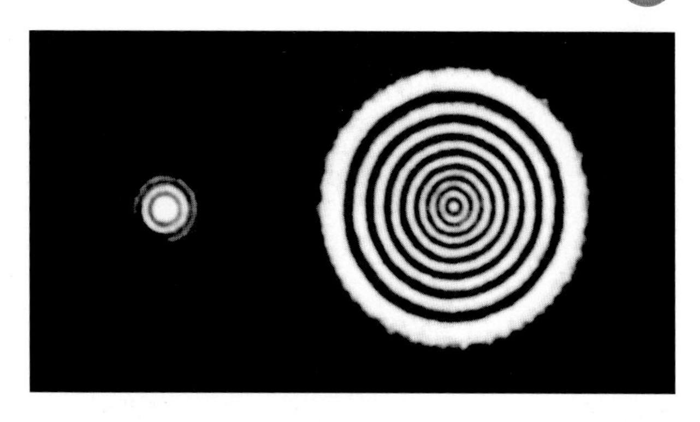

Figure 3.13. The classic view of the Airy disc of a star seen at focus in a well-corrected and collimated unobstructed telescope (left); at right is the even, symmetrical Fresnel diffraction pattern as seen in the same system when racked far out of focus.

achromat will focus within a range of about 1/2000 of the focal length, and with the same degree of magnification, creating a crisp and accurate image.[23]

Viewing the image of a white star when the atmosphere is calm, the user of an evenly figured achromat of this type, well corrected for spherical aberration, will see the Airy disc at focus, surrounded by one bright diffraction ring, and a much dimmer one. Circular targets of expanding rings (diffraction or "Fresnel" rings) appear when the telescope is racked inside or outside focus. Their central discs and outer rings should have nearly identical proportions and intensities. The colors will be different, however. With good, standard color-corrections the image of the star expanded inside the focus – moving the focuser toward the objective – will show a green center with purplish outer rings. Outside of focus, the center will turn purple, while the outer rings go greenish. Some manuals use only the tint of the outer rings to describe the general color.

Semi-Apochromats and Apochromats Using standard crown and flint glasses in a good achromatic design like Fraunhofer's, a purplish haze of defocused, uncorrected light is evident around bright objects. It can't be further reduced without changing glass types and the curves and thicknesses of the lenses. Later designers used heavy "short" flints for better color corrections, and have now gone to extra-low dispersion glasses (ED glasses) for the same general effects. Sophisticated objectives of this type focus all colors within a range of about 1/4000 of the focal length. Such a semi-Apochromat creates an extremely

---

[23] Opticians making a two-lens objective of these "normal" crown and flint-type glasses generally attempt to correct the focus for colored light at two of the solar spectral lines of hydrogen discovered and named by Joseph Fraunhofer (1787–1826), namely the "C" line (red, at a wavelength of 6562.8 Ångstroms) and the "F" line (in blue, at 4861.4 Å). The green light, at around 5600Å, focuses slightly ahead of this crossing point An image so focused brackets the range of the eye's greatest sensitivity in the yellow-green range, creating a satisfactory, realistic image.

Figure 3.14. This 70-mm aperture semi-Apochromat refractor, a Tele Vue Pronto model, exemplifies the small, fast (f/6.8) new generation of refractors for learners and serious buffs that require a portable glass for air travel and photography. The Gibraltar-model altazimuth mounting is steady and intuitive to use.

clean image at focus. The term was coined by the Zeiss firm to describe certain of its objectives, mainly the "AS" and "B" types. Unless the viewer defocuses slightly to check for remaining stray color, most cannot tell the image quality from that rendered by the highest type of objective, called the Apochromat or "Apo."

Apochromats

In sports, once a "faster runner" comes along, the cognoscenti immediately raise the bar. With optics, on the other hand, excellence is relative and opticians may doggedly defend the regime they learned under. Consider the excellent Taylor–Cooke triplet objective as an example. The "Cooke" was long hailed as "Perfectly Achromatic"[24] or "Apochromatic," although newer types made with

[24] Taylor, p.78 – see the Bibliography.

exotic materials had soon actually beaten its level of performance for color correction and optical aberrations.

Today's optician designs an Apochromat using a combination of well-matched flint or crown with ED glass, sometimes a fluoridated type, or by using the actual mineral crystal Calcium Fluorite as an element, combined with matched special glasses. The best modern Apochromats have three or more lenses in the objective. The standard set by objectives such as the Zeiss APQ Fluorite type has not yet been surpassed at the time of this writing. From interferometric test data available, it can focus the primary colors at nearly a plane surface, the focal region having the depth of 1/16,000 the focal length, or better, over a wide range of "hues" or wavelengths.

In the 1960s, legendary optician James Baker defined a class of ideal instrument he called "superachromats" in the attempt to create a little more headroom for optical definition. The term never caught on. We still call the best refracting objectives Apochromats, leaving quite a bit of wiggle-room for maker-definition of quality, since there is no "Underwriter's Laboratory" for commercial optics. Few match the Zeiss standard, yet most that are somewhat better than a "semi" simply acquire the Apo label.

With an Apo that has coma and spherical aberration fully corrected, the center and rings of the Fresnel images both inside and outside focus reveal the pure color of the star under observation, without colored fringes or deviations in ring and center widths. This is a standard attained by only a few industrial makers, among them Zeiss, Tele Vue and Astro-Physics in the United States, and Takahashi and Vixen in Japan. New makers are on the horizon, however, and true Apo refractors may be somewhat less expensive and easier to obtain in the future.

In most two-lens objective designs, the weather-resistant lens, the "crown" lens, faces out, while the more vulnerable "flint" lens or other exotic glass with high mineral content rests behind it, inside the cell. This is not always true, however, since designers of the past such as Steinheil and Clark often placed the flint element "forward." Makers of Apos endeavor to protect the exotic or fluorite elements of their objectives, placing them between other lenses, or behind a less sensitive front element. All refractors require care in protection from thermal shock.

Reflecting and Compound Telescopes

Newtonian Reflectors The light that forms the image in a Newtonian reflector passes from the sky, down the tube to the primary mirror at the bottom. It reflects back to the mirror's focus point, converging onto the surface of a small, flat secondary mirror placed near the upper tube opening (see the illustration in Chapter 10). The reflecting surface of the secondary, set at a 45-degree angle to the optical axis of the primary, turns the converging light 90 degrees, sending it to the side of the tube to focus the image for an eyepiece. Although older designs often incorporated square or octagonal flat mirrors or even prisms, most "secondaries" are now elliptical mirrors. This is the most optically efficient shape, since the tilted mirror presents a circular silhouette to the incoming light beam, causing the least disturbance or diffraction of the image.

Long-focus Newtonians: Reflectors with focal lengths six or seven times the mirror diameter or longer, regarded as the classical instrument of all work by many amateur observers, have a comparatively wide coma-free field. Unfortunately, equatorially mounted Newts of f/6 or f/7 with more than 10-inch (25.4 cm) aperture are a disappearing breed, outside of permanent backyard and garden installations. The size and weight of the mount, pier, and counterweights make them lumbering dinosaurs in an era of increasing mobility and dark sky

Figure 3.15. Pier-mounted Newtonians like this sidereal-drive 12.5-inch (32cm) f/6 Research-grade Meade set the standard for late 20th-century advanced amateurs. Their size and bulk confine them chiefly to permanent installations. Observers currently favor large Dobsonians for their ease of transport to dark-sky sites.

observation populated by larger, shorter and lighter-weight alt-azimuth Newtonians.

Aperture being equal, an optimized Newtonian system with excellent mirrors can give nearly refractor-like images at f/ratios of 10 and above. A personal homemade favorite of the type is a 3.4-inch (86 mm) f/14.5 equatorial fabricated from 1960s-vintage Edmund Scientific optical parts. The tube is PVC pipe mounted on aluminum. I fabricated the entire mounting and slow motions from a bicycle frame and parts, after coming across the notion in an early book by Sir Patrick Moore (see illustration).

However, we live in a world of what can be readily obtained. Although one may custom-order any length of system one wants, the price can be prohibitive. Commercial suppliers have ceased to produce large-aperture optical tubes of unwieldy dimensions for the amateur market. After all, a properly made 8-inch

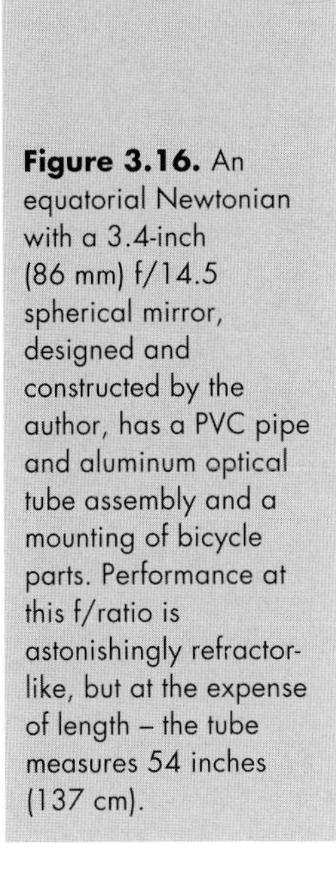

Figure 3.16. An equatorial Newtonian with a 3.4-inch (86 mm) f/14.5 spherical mirror, designed and constructed by the author, has a PVC pipe and aluminum optical tube assembly and a mounting of bicycle parts. Performance at this f/ratio is astonishingly refractor-like, but at the expense of length – the tube measures 54 inches (137 cm).

f/10 tube is more than two meters long, with proportional weight! This is one practical reason why Brachytes and other "folded" systems are popular among amateur telescope makers who want the optical advantages of a long-focus reflecting system.

Dobson Newtonians: Fast, modular Newtonians on trunnion-and-turntable mountings, usually called "Dobsonians" after the U.S. innovator John Dobson (1915–), are the instrument now most often recommended as a practical and cost-effective solution for the observer who wants affordable large aperture. The increased light grasp is a great advantage to the visual deep-sky enthusiast. In fact, the advent of the type is arguably a watershed in the history of visual astronomy. Sheer aperture gives big Dobs such visual punch that observers can easily acquire and examine dim galaxies and features of nebulae that earlier observers never dreamed of seeing in an amateur-class scope.

Most large Dobsonians have primary mirrors with focal ratios in the neighborhood of f/4.5. Dobs that support the upper secondary-mirror cage with truss poles are easy to disassemble, and break down for easy transport in a common auto. Newer types of mirror cells, rigid locking truss-tubes and ultra-lightweight upper cages have largely overcome the problems of "tube" flexure that once plagued collimation in such designs.

Performance

Even the average Dobsonian telescope is excellent when used at low to medium magnifications; say up to 30× the aperture in inches. The observer can also appreciably increase the coma-free field of a fast system by using an available lens-type corrector at the focal region, much as is done with large, fast professional telescopes.[25] The buyer should be aware, however, that only models in the higher price ranges are up to the challenge of viewing extremely fine celestial details under most conditions.

Two principal factors operate here, both of which require design time and expert artisanship to achieve. First, the figuring and parabolization of the thin mirrors used in lightweight larger aperture Dobs is an art that doesn't lend itself to current methods of mass-manufacture. This is why you will often see the name of a custom mirror-maker noted in descriptions of large, high-quality Dobsonians. Also, thin mirrors over 10-inch (25.4 cm) diameter require specially designed and fitted mountings, to avoid the mirror flexure that deteriorates images when the Dob is moved to certain viewing angles above the horizon. New methods of mechanically warping a mirror to perfect its figure during observation show promise, but as of this writing are still in the bailiwick of expert amateur makers and observers.

With mirrors up to around 16 inches (40 cm), mounting in a low-profile housing keeps the eyepiece close enough to the ground for comfortable viewing from a standing position, or with a small ladder or platform. The accompanying

---

[25] Tele Vue's ParaCorr™ was developed for Newtonians with 2-inch (50.8 mm) focusers.

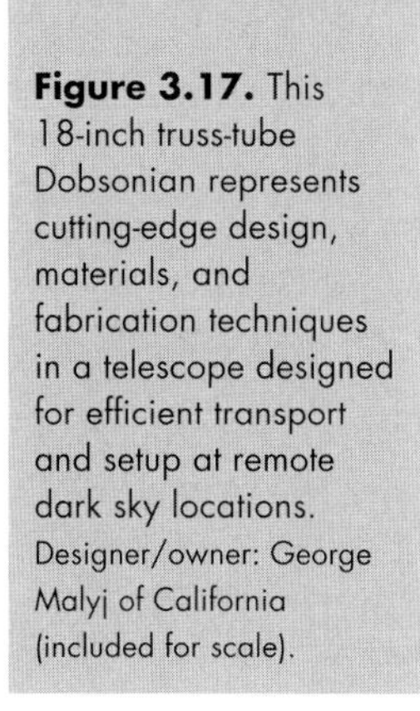

Figure 3.17. This 18-inch truss-tube Dobsonian represents cutting-edge design, materials, and fabrication techniques in a telescope designed for efficient transport and setup at remote dark sky locations. Designer/owner: George Malyj of California (included for scale).

illustration shows an excellent example of a home-designed and constructed portable 18-inch truss-tube Dobsonian.

Tracking with a Dobsonian: Although some have successfully used the manual Dobsonian configuration for sketching and limited bright-object photography, applications beyond visual observation and searching the sky by sweeping require a more complex mounting than the manual rocker-box. There have been almost as many designs as makers. One of the earliest was the Poncet platform drive, which utilizes a fixed pinion for the south-facing end of a flat mounting plate, with the "front" or North end of the platform driven along an inclined plane, imparting a close approximation of Sidereal driving motion. Often, a lead-screw turned in a trapped nut at the proper rate by a DC motor provides the tracking, by driving blocks faced with Teflon® or other slick synthetic material mounted to underside of the azimuth platform. Makers have developed various ingenious solutions for "resetting" such drives, most of which have a relatively short period of accurate tracking.

Digital altazimuth drives: Based on professional models, these soon entered the picture, with software-controlled stepping motors attached to the altitude and azimuth axes. Original versions, often controlled by clunky desktop computers, have recently been mainstreamed as factory-installed units or add-ons, using ROM chips and integrated control boxes and driving motors. Much of this progress was driven by the success of commercial computer-driven fork mounted SCTs, and by privately designed open-source versions, with plans provided by dedicated observers worldwide, such as Mel Bartels of Oregon.[26]

---

[26] Driving Dobsonian telescopes is described in detail in books such as Texereau (2nd Edition, Appendix I), and Kriege and Berry's *The Dobsonian Telescope* – see the Bibliography.

Compound and Catadioptric Systems After Galileo demonstrated Copernicus' idea of the Sun-centered Solar System through optical means, telescopes quickly caught on as a new test-bed for natural philosophy. The optical texts of Alhazen, Witel, Kepler, Huygens and Descartes became pedagogical staples, and the acquisition of a perspicillum or "Dutch tube" became a requisite for all learned folk. Tracts linking optics with natural philosophy flowed from the letterpresses. Literate 17th-century monks had much to offer on the subject. French clerics Marin Mersenne (1636) and Guillaume Cassegrain (ca. 1670) concocted feasible designs for compound reflecting telescopes that are still in use today, although technology wasn't up to creating the mirrors required until centuries later.

Gregorian: Meanwhile, a Scotsman had designed the first compound reflector to see general use. Presented by James Gregory in his Optica Promota of 1663, the Gregorian is laid out like the Cassegrain (see below), and has an elliptically figured secondary mirror placed beyond the focus of the primary to collect and send the light beam back through a perforated primary mirror. The twice-inverted view is upright and correct, as in a terrestrial spotting scope or binocular. As with Mersenne and Cassegrain, contemporary opticians couldn't produce the necessary surfaces, but the Gregorian saw light in the mid-18th century under the hands of several workers, including the British optician Short. Although the fields of view in such instruments are extremely narrow by modern standards, many naturalists and amateur scientists used Gregorians through the mid-1900s as a false-color-free alternative to the refractor. Today, the giant radio telescope at Arecibo uses a Gregorian system to feed signals from its sensors, and the design has been used for special applications in Earth-orbiting telescopes.

"Classical" Cassegrains: In this design, light striking a fast paraboloidal, perforated primary mirror converges to a secondary mirror of hyperboloidal figure, which relays the light back down the tube, through the aperture in the primary to focus. Ideally, the maker provides the system with a (conical) baffle tube to exclude stray light. The Cassegrain focus is beyond the mirror-mounting cell, readily accessible to an eyepiece, camera, or other instrumentation. This is not the only mirror-set usable, and the term Cassegrain is actually generic for any system with a concave primary mirror and a convex secondary in line with it. Fabricating the two mirror surfaces to several other complementary figures produces optically sound results.[27] This focal position is always identified as "Cassegrainian," although it was Mersenne who first grasped the possibility, as noted above.

The focal ratio of a Cassegrain system is generally "slow," about f/14 to f/20, giving high natural magnification and a narrower field of view than conventional Newtonians. The folding and amplification by the secondary, however, allows a

---

[27] The difficulty of producing a hyperboloidal secondary can be avoided by leaving it spherical and under-correcting (aspherizing) the figure of the primary. This is essentially the Dall–Kirkham solution for which Thompson, p.7, references *Scientific American*, June 1938. There is also a useful table in Rutten and Van Venrooij, p. 65 – see the Bibliography.

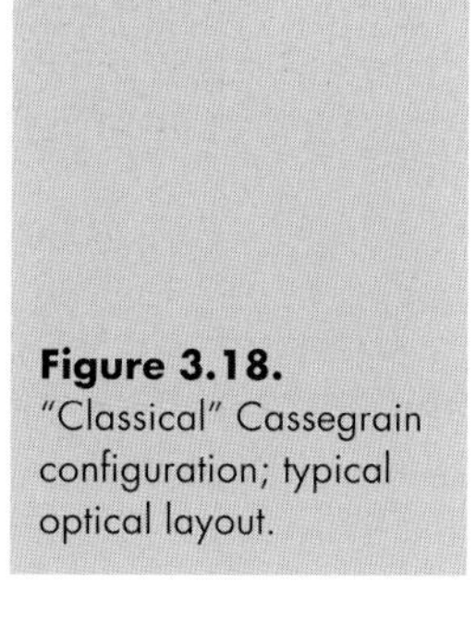

Figure 3.18. "Classical" Cassegrain configuration; typical optical layout.

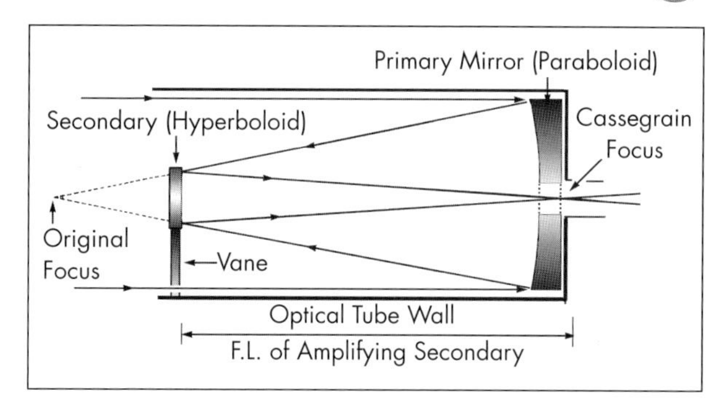

much shorter tube (see illustration). More popular with amateur astronomers in earlier decades, it has now picked up the rubric "classical" Cassegrain to distinguish it from the Schmidt–Cassegrain. Cassegrain systems are still available, but not often encountered in the mass market.

Schmidt–Cassegrain: This is the most popular member of the family of compound instrument known as *catadioptric* (lens–mirror) or "cat" telescopes. The "Schmidt–Cass" or "SCT" is a result of combining the basic mechanical arrangement of the classical Cassegrain with a wide-field camera designed in 1929 by the Estonian-born amateur optician Bernhard Schmidt (1879–1935). The prolific U.S. optical designer and engineer James G. Baker envisioned this hybrid in the 1950's. It uses three optical elements to bring the image to a Cassegrain focus.

The "Schmidt" portion of the optical design is the front *corrector plate*, a plano-concave lens of toroidal figure that conditions the incoming light to nullify the spherical aberration imparted by the second element, a fast spherical primary mirror of around f/2. The corrected light converges from the primary to strike a convex, spherical secondary mirror mounted in a cell on the backside of the corrector plate. The secondary amplifies the focal ratio by about 5×, sending the corrected light cone back through a sturdy light-baffle tube assembly mounted on the rear cell of the instrument.

Unlike the classical Cassegrain, the SCT primary mirror moves to focus in standard commercial systems. The baffle tube assembly must be sturdy, since it also supports the primary mirror and focusing arrangements. Mirror-focus allows convenient use of accessories requiring considerable alteration of the focal plane position. Due to its short length and balancing qualities, the optical tube design is also particularly useful for imaging. Using a rack-and-pinion or Crayford friction-bearing focuser on the rear cell, the user can keep the primary mirror in optimal position and also achieve precise focus control for high power viewing, CCD work and astrophotography.

The original SCT of commerce was a design by Tom Johnson of Celestron International. The optical tube was mounted on a compact dual-fork, having an AC synchronous motor in the base that drove the unit to follow the stars at the Sidereal rate. They were soon available on German equatorial mounts as well. Celestron's SCTs usually yield around f/10 or f/11 at the Cassegrain focus, and f/6.3 versions are available from Meade Instruments, which has been making Schmidt–Cassegrains of its own design for decades. Both makers provide focal reducing and field-correcting lenses that quicken the focal ratio to f/6.3 or faster. If a novice wants to start a lively discussion amongst amateurs, a sure-fire way is to raise the question of which firm makes the better SCT. The ensuing earful will be an education in itself, since a friendly rivalry between owners has long been a hallmark of the observing scene. A few additional makers worldwide supply the balance of SCTs at this time.

The SCT was the first amateur-class instrument made widely available with motorized computer-controlled tracking and pointing, now available on most models.[28]

Maksutov: Another type of "cat" telescope folds the light path, based on the designs of Bouwers and Maksutov from the 1940s. Bouwers apparently patented the design first under wartime conditions of secrecy in the Netherlands, but Maksutov was first to publish it, conferring eponymy on the "Mak."[29] We get into specialist territory here. Depending upon the variant (and there are several), this design replaces the secondary mirror of the Cassegrain with a silvered mirror spot coated on or attached to the reverse of a thick, spherical meniscus corrector plate mounted at the front aperture of the tube. Bouwers' original telescopes had an achromatic (two element) meniscus corrector to control spherical aberration, while Maksutov employed different curvatures on the corrector's surfaces for the same purpose. The nominal optimal focal ratio is around f/15, making it one of the slower telescope designs.

The type has excellent visual and photographic qualities and a flat field when properly constructed, comparing favorably with the traditional long-focus refractor or Apochromat. The Gregory–Maksutov variant is a development by Texas optician John Gregory, taken up by the Maksutov Society decades ago. It proved a great favorite among US amateur telescope makers, who have produced many fine examples.

Due to the thickness of the corrector, Maksutovs are generally heavier than SCTs of the same aperture, and more expensive to construct. Due to the mass of the optical elements, Maks take a while to reach thermal equilibrium when the temperature drops. Some makers supply units with "zero-expansion" mirror materials to decrease this factor, pioneer maker Questar being among the earliest to do so. Other design features have been developed to cope with thermal factors, such as the removable primary mirror cover on units designed by Roland Christen of Astro-Physics. The Maksutov is currently available in apertures from 2.4 to 12 inches (60 to 305 mm) from various makers worldwide.

Maksutov–Newtonians ("Mak-Newts") and *Schmidt–Newtonians* ("SNTs") are additional hybrid variations on the "cat" scope that have recently gained new prominence through production by several manufacturers and custom makers.

---

[28] The author constructed an analog motor-slewed fork-mounted SCT in 1987, with slewing speed controlled by rheostats. The positions were simply read off illuminated setting circles as the mount slewed to them. This was not arc-second accurate, but sufficed for locating bright objects rather quickly, since the slewing speed was infinitely adjustable between 16× Sidereal and a fast-slew on both axes of 5 seconds from horizon to horizon.

[29] According to Fillmore's Preface to *Construction of a Maksutov Telescope* – see the Bibliography.

They are gaining great popularity among observers. The Mak-Newt is a minimally obstructed system that has gained a reputation for excellent visual qualities, while most users prize the SNT primarily for imaging, due to its fast optics and easy portability.

Unobstructed Reflectors Reflecting systems are free of color error, and the contrast gain due to eliminating the central obstruction makes such designs a favorite alternative to the refractor for practical observation work. Designing and creating systems with image aberrations controlled has been an irresistible challenge to many opticians. Designs are generally long-focus "folded" mirror systems that can be figured and tested with variations or extensions of techniques already familiar to the practiced amateur or professional "glass pusher."

Herschelian: This is the simplest "unobstructed" type, named after discoverer of Uranus, German–English musician–composer and telescope maker Sir William Herschel (1738–1822). Herschel arrived at the form through ingenuity. His larger speculum-mirror telescopes had no Newtonian secondary mirror; the observer looked into an eyepiece pointed at the primary mirror from the edge of the optical tube. Herschel tilted the primary mirror slightly to focus its beam at that point. The oversized tube, large aperture and long focal length made any slight obstruction by the observer's head almost inconsequential (see illustration).

A workable Herschelian can be made from any regular Newtonian by (a) physically masking off all but a circular section of a Newtonian primary with half the primary's diameter, (b) removing the secondary mirror and spider and (c) slightly tilting the primary using the collimation bolts. Moving the primary forward or making an angled cutaway at the "focus" side provides room for the exit beam, with the focuser mounted to the tube exterior. This introduces significant optical errors but the experiment creates a satisfactory image for some purposes.

In the modern version made from scratch, the optician figures an *off-axis paraboloidal* primary for the purpose, or cuts the blank from the perimeter of a larger, pre-figured paraboloidal mirror before the reflective coating is applied. This mechanically offsets the optical axis, reducing astigmatism and other aber-

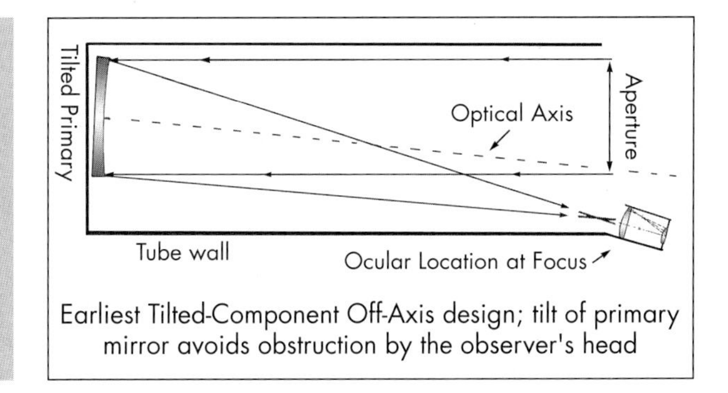

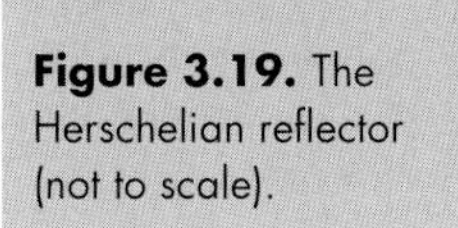
Figure 3.19. The Herschelian reflector (not to scale).

rations due to the reduced tilt of the primary. Properly done, it also eliminates obstruction from the eyepiece holder and the observer's head.

There are no current makers of this type for the amateur market. Off-axis paraboloidal mirrors designed for industrial collimators are, however, available at premium cost from a few custom suppliers. For those interested in new development of this classic type, an extensive discussion of an off-axis paraboloidal Herschelian with a bowed-spar truss optical tube appears in the recently republished two-part article by Gary Frishkorn.[30]

Brachytes, Schiefspieglers, and other TCTs: There are many other unobstructed reflecting designs, generally lumped under the rubric of Tilted Component Telescopes or TCTs. The challenge is making a compact optical assembly while controlling coma, astigmatism and other image-plane aberrations arising from the necessary tilting of the components to avoid obstruction. Two-mirror folded "Brachyte" reflectors or "arm telescopes" lie at the simpler end of this broad class of instruments. The *Schiefspiegler* ("leaning mirror") designs by Anton Kutter of Germany, and variations on the "Yolo" design introduced by Jose Sasian of California in the 1980s, are more complicated. Wolter, Francis, and others such as Peter Hirtle of Seattle have recently innovated and sometimes published complex folded designs.[31] TCTs may incorporate secondary and tertiary mirrors figured as toroids, or a to variety of conic sections from hyperboloids to ellipsoids. Some designs use aspheric or wedge-figured correcting lenses, prisms, or employ mechanical warping of optical elements to achieve the necessary corrections.[32]

Although technically marvelous, such telescopes are difficult to design and fabricate. They have (so far) tempted no manufacturer to begin production, and remain the province of custom opticians and skilled amateur telescope makers (see *A Note on Tilted Component Telescopes*).

Siderostats and Fixed-Eyepiece Telescopes

Like the TCT, these instruments – usually installations, really – are difficult to design and construct. They are generally either custom-made or fall into the province of the amateur opticians and mechanician. A fixed eyepiece position receives the focused beam relayed from the primary optic through an optical tube or axis assembly. The aperture is commonly fed light by a moving flat mirror with its rotational axis pointed to the Celestial Pole.

---

[30] *The Best of Amateur Telescope Making Journal*, vol. 2, Sections 13-4 and 14-3, – see the Bibliography.

[31] See: See Gary Frishkorn, "Design and Construction of a Modern Herschelian;" Robert Novack, "A General-Purpose Yolo You Can Build;" John Francis, "Improved Test Methods for Elliptical and Spherical TCT Mirrors," and Heino Wolter "The Multi-Schiefspiegler" in *Best of Amateur Telescope Making Journal*, v.2 – see the Bibliography.

[32] Controlled thermal- or vacuum-warping of the mirrors in Newtonian configurations to improve the figure is also one of the newer experimental practices of advanced amateurs.

Figure 3.20. An unusual fixed-eyepiece equatorial mounting that uses mirrors to bring the focal plane to a central position on the polar axis. The 6-inch refractor's optical tube is counterweighted to maintain balance, as in a Springfield mount. Design and construction: Thomas Dobbins.

A Polar Siderostat has a single mirror that reflects the light from a chosen part of the sky into an objective or primary mirror. The French natural scientist Boffat first introduced the concept late in the 17th century. Others employed the concept for various telescopes around Europe, and Foucault ingeniously refined the mirror mount in 1869 for use with horizontal telescopes. Due to geometric relationships, the image slowly rotates around the center of the field of view as the mirror tracks.

Coelostat: This "sky stopping" type of mount conquers field rotation, either by moving the telescope axis itself in Declination as the driven flat mirror tracks parallel to the plane of the Earth's axis in Right Ascension, or by using a second flat mirror to compensate. The 150-foot Solar Tower telescope at Mount Wilson is a refinement of this type that directs light through a 12-inch Hastings triplet-Apochromat objective to a subterranean instrument room. It images at many wavelengths, and keeps track of sunspot movement in real time by daily hand-annotated tracings from its ~42-cm-diameter projected white-light solar image.

Several optical firms offer custom siderostats and related designs for specialized clients, usually educational institutions, planetariums, or research facilities. These sometimes require a roof-mounted siderostat or coelostat to feed a solar beam to a stable instrument platform or demonstration apparatus. Solar telescopes of this type are called *heliostats*, from the Greek for "Sun stopper." Many amateur telescope makers, including the author, have constructed various forms of siderostat. The mirror housing assembly of the author's instrument, a portable Boffat-type (top mirror) Polar Siderostat, is illustrated (Fig. 3.21).

Professional designs for instruments with heavy equipment in place often employ a fixed Coudé ("elbow" or "folded") focus. The classic type is due to Loewy of the Paris Observatory; other designs followed. Loewy's original Coudé style employed an objective lens housing that slews in Declination and directs light through flat mirrors or prisms down a hollow polar shaft to the focal point.

Figure 3.21. The mirror assembly of a portable 70-mm f/11 Polar Siderostat with Solar rate AC drive; the fork-mounted elliptical mirror slews in declination, the entire canister housing slews in right ascension. The projecting silver arm is a dynamical counterbalance. Design and construction: The author.

The 6-inch (150 mm) Zeiss instrument illustrated is a modern version of this type. The term, like "Cassegrain," has become generic to a remote focal position where flat mirrors and/or lenses direct the beam by folding, from the primary optic through the polar (or azimuth) axis to the observing station.

Another fixed-eyepiece mounting favored by mid-20th century ATMs is the Springfield. This style uses prisms or flat mirrors to bring the focal plane to an observing station near the secondary mirror position of a counterbalanced Newtonian. James Muirden states that it is one of the "mounts to avoid," as the image is inverted in an unusual way. I have only seen one in twenty years, at a place called Stellafane in Vermont (where there seemed to be a good many other odd telescopes about). We may safely infer that the writer need have feared no stampede toward constructing the type![33]

Point-of-Purchase Checks

The following will be helpful in sorting out the bad apples when you begin looking for your first telescope.

Checkpoint List

1. Advertising hyperbole: Don't let performance claims and glowing testimonials in ads and on packaging mislead you. Take glib statements by sales per-

[33] "The reasoning behind the Springfield may be sound, but the lack of published observations made with it suggests that it is the plaything of the amateur optician and mechanic rather than useful to the practical astronomer." *The Amateur Astronomer's Handbook*, p. 49 – see the Bibliography.

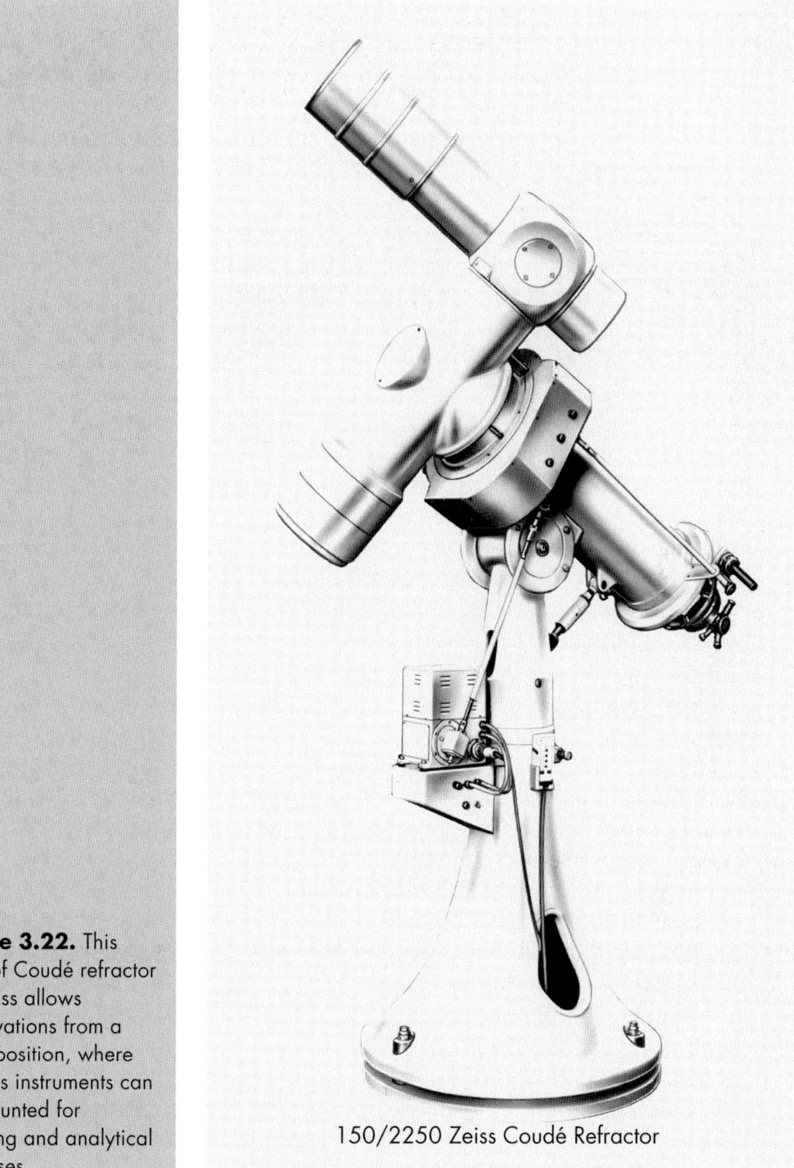

Figure 3.22. This style of Coudé refractor by Zeiss allows observations from a fixed position, where various instruments can be mounted for imaging and analytical purposes.

sonnel with a pinch of salt. Claims of "high power" (or a fantastic power range) are often the main sales pitch. As we know, the choice of eyepiece determines the magnification for the telescope. The *quality* of the optics and the useable *surface area* of the lens or mirror are the real criteria. As Terence Dickinson expresses it in his excellent *Backyard Astronomer's Guide*: "... the maximum useable power for a 60-mm telescope is only 120×. Claims that such a telescope can magnify 400× are misleading, intended solely to lure the unsuspecting buyer." In any case, experienced observers with good eyesight generally use modest magnifications that will clearly define the object of study.

2. If there is no display model available and the sales manager will not allow you to examine a telescope out of the packaging, don't risk the purchase; go elsewhere.
3. Look closely for good workmanship and sturdiness, paying attention to the fit and finish of components. In most climates an astronomical telescope spends many hours covered with nighttime dew, so examine the painted or anodized surfaces and all plated metals for corrosion-prone skips and blemishes.
4. As of this writing, except for the occasional tube (or truss set) made of carbon-graphite or similar technical composites, good optical tubes, tailpieces and cells for mirrors and objective lenses are constructed of metal and alloys: generally aluminum. Check the main tube, the finder, the focusing tube and track, as well as the "star diagonal" supplied with a refractor. Also, check the eyepiece heads for injection-mould marks that indicate plastic parts. Plastics, if used at all, are reserved for incidentals such as knobs, pads and cover plates. Another exception is the hard plastic or composite shrouds used to cover the motor assemblies and electronics of "go-to" telescopes of the fork-mounted type. These are covers, not structural members. The framework underneath is usually of cast and machined alloy.
5. A simple, steady mount is better than a shaky one fitted out with impressive gears and knobs. Examine the focusing tube, mounting block or cradle, tripod head and all other moving parts of the mount and optical tube. Check for obvious looseness that you can't eliminate by tightening bolts or knobs (which should be readily accessible and adjustable without special tools). Viewing, even at low magnification, will grossly magnify seemingly minor "wobble" in these parts, and make observing nearly impossible at higher powers. If the mount has slow-motion controls for tracking moving objects, try them in both directions. These should actuate a smooth sweep without annoying slack or "backlash."
6. While looking this closely, one might as well check the lens or mirror surfaces as well.

   Refractors: With a refractor, take out the eyepiece and look into the tube while holding the lens up to a light source. This grossly exaggerates any dust particles present; it is easy to tell if the assembly is clean or not. Dust on the outside of the lens is easy to deal with, and a light dust coating on the outer surfaces of display models is to be expected. Foreign material between the lenses or inside the objective, however, is clear ground for a "no sale" verdict, unless the outlet will allow you to inspect a model freshly out of the box. If the fresh model has the same faults, pass, no matter how attractive the price.

   Newtonians: To inspect a Newtonian reflector, take out any eyepiece and inspect both the diagonal and primary mirrors while pointing the tube toward a light source. More than a light dust coating, or a mottled appearance of the mirror coatings is a clear call to ask to inspect a unit from a sealed box. As with a refractor, if such irregularities recur, pass on the purchase.

Make similar checks for interior contamination or poor coatings on the mirrors of closed-tube compound designs such as Schmidt–Cassegrain telescopes (SCTs), Maksutovs (commonly known as a "Maks"), Maksutov–Newtonians ("Mak-Newts"), or Schmidt–Newtonians ("SNTs"). Inspecting with bright lighting, you will detect at least a few flecks of dust or particles on the mirrors or the backside of the corrector plate. This is normal, and it won't affect image quality.[34] A mottled appearance of the coating surface, or light scratches ("sleeks") on the mirror are another thing altogether, and cause for rejection or return on more than cosmetic grounds.

A further check is quite useful: the cleanliness and coating condition of the secondary mirror and the crucial baffle-tube areas in both SCTs and Maks. Remove any rear cover or diagonal attachment to get a good look up the baffle tube at the coating of the secondary mirror and the interior of the tube. A manufacturer once released a batch of Schmidt–Cassegrains that had shiny interiors in the light-baffle tubes, causing noticeably lowered contrast and glare from off-axis light, compared with previous production runs of the scope. Dealers detected the flaw, returned units, and the factory made good in the next production run. "Lemons" get through quality control in any industry, and such flaws indicate shoddy assembly of the particular unit. They may also be an indication of substandard performance of the model itself, warranting further checking of its track record via test reports and interest group commentaries before purchase.

7. Ask to perform a cursory daytime image test with the store display model. Again, if this courtesy is refused, you should strongly consider purchasing elsewhere. The testing will take only a few minutes, and it is well worth your time to do it.

   Most display models of refractor or SCT will have a right angle viewer such as a star diagonal ("zenith prism") or terrestrial prism unit mounted on the focusing tube, holding the eyepiece. This, of course, is not a factor with Newtonians. A mediocre diagonal will compromise the performance of a good lens, so try removing it before testing. Fit the longest eyepiece supplied (usually about 20 to 30 mm focal length) into the focusing tube, and rack or slide it out to reach focus. This may prove impossible to do. If so, replace the diagonal.[35]

   With or without a diagonal in place: if you wear eyeglasses, remove them for the time being. The optical system will compensate for all but severe

---

[34] Many amateurs lacking a true clean-room have tried and failed to clean such optical tube interiors to better than "factory spec." Should significant buildup occur, shipment to the manufacturer for professional cleaning is the most practical solution.

[35] There will always be situations where you want to observe straight through. The lack of a supplied extension tube, a secondary drawtube (rare in today's designs), or a focuser that can't be racked out far enough to compensate for removing the diagonal, all indicate poor design. This is, unfortunately, common in most commercial models, even many with first-class optics and mountings. Most manufacturers simply assume that their provisions are adequate. Other, more prescient designers add fixed-length extension tubes to their accessory line. These are adequate, but do not allow the same flexibility in focal range as a drawtube, and are usually not standard equipment.

astigmatism in the eye, and it is usually difficult to obtain the full field of view wearing glasses. Focus the telescope on a high-contrast object ten meters or more distant. A white signboard with dark letters or any dark outline against a white background will serve.

If edgings of bright color appear along the dark/light boundaries of what you are viewing; if the perimeter or other portions of the field of view appear fuzzy while the rest is in focus, or if straight edges crossing the edge of the field of view appear strongly or asymmetrically curved, avoid the instrument. These are all clear indications of design or manufacture problems.

Check the image for sharpness by examining a sample of small print. Half-inch (1 cm) lettering should appear well formed and perfectly clear at this distance, with no "phantom" shading between letters or lines. Casual daytime testing won't let you decide astronomical quality, but an instrument that is noticeably deficient in the daytime is bound to be a grand failure under the greater demands of the night sky.

For simple economic reasons, many manufacturers concentrate on the main optics, supplying otherwise sound telescopes with eyepieces that are variations of the inexpensive Kellner or, more recently, adequate but rather generic symmetrical Plössl types. Should you decide to purchase the instrument, you can always upgrade.

8. Return Policies: After purchase, you may find that a particular model of telescope doesn't suit your requirements. Discuss the fine points of the supplier's return policy *before* you purchase. Return after a nighttime trial may be impossible unless there is an obvious mechanical defect. Even in that case, returns may be limited to repair or an exchange for an identical model. Checking return policies is, obviously, even more important when buying out-of-area through mail or Internet order. The buyer customarily pays full shipping on factory returns for service, a sizeable sum with a large or heavy instrument.
9. Mail-order by catalog or Internet: You may find your dream telescope offered at a very competitive price through an Internet or catalog mail order outlet. Just beware of illustrations showing accessories that are <u>not</u> included at the package price. With an honest advertiser, a look at the fine print will reveal the facts, so carefully go over the descriptions of included features and accessories. Some mainstream manufacturers consider even crucial items (tripods, wedges, finders, drive motors) to be add-ons. Carefully examine the descriptions of accessory packages as well. Obsolescent eyepiece models, eyepiece filters in seldom-used colors, unsupported pointing software versions and the like, make such packages no bargain.
10. "Deep discounts": Pricing far below the median may legitimately occur from overstocking, when a model is dropping from production, or an item is used for demonstration. On the other hand, unauthorized dealers sometimes pop up selling second-quality optical goods, "back door specials," designs unsuitable for astronomy, and instruments stripped of accessories at steep discounts. Stolen instruments also occasionally surface in this way. *Never* purchase a telescope sight-unseen off the Internet or from a print ad unless there is a physical location listed, and a telephone number you can call first for information.

11. Custom instruments: These may require long waiting periods, especially the larger Dobsonians with premium quality hand-figured mirrors, and Apochromatic refractors of the highest rank. You can expect to pay top dollar on the used market as well, since sellers are aware of the demand and scarcity of many models. Bearing in mind that good optics such as these are usually worth waiting for, lack of speedy production is usually a sign of care, not carelessness.

    Special additional checks for refractors: see Chapter 2 under *Used Equipment*. You may not be able to do these before acquisition, but they form a practical basis of evaluation for new refractors on Warranty.
12. Finally, as noted earlier, waiting times up to six months or longer are increasingly common for new models on the market, especially those pre-announced by the larger factories. To avoid prohibitive delivery times, call or e-mail ahead of time to insure that the item is in production, in stock, or available by special order on a timetable. Otherwise, you may find yourself at the end of a long list of unfilled orders.

Shipping and Warranties

There are many consumer issues around telescopes and equipment, just as in any field. There are a few points of difference worth mentioning, so here's the "short list" from experience on both sides of the fence.

Shipping Costs and Modes

When comparing prices, be sure to get an exact full quote on all shipping and handling costs for items, plus any miscellaneous charges for the order itself. Some better companies effectively subsidize part of their shipping cost to get your business, typically adding a low shipping charge of a few currency units on a sliding scale. This can be a very good deal. On the other hand, some "deep discount" dealers count on profit from so-called "handling fees," often a flat charge, and then add a per-item charge for shipping as well. You may find ordering from a local outlet is a better deal.

Transatlantic orders will always cost considerably more than continental orders. A typical fee for a small item sent by International Air Mail at this time is a minimum of about 15 U.S. Dollars (14 Euros). If the company has no regular import business, you may also find yourself paying a customs brokerage fee and waiting for the item to clear Customs before you can take delivery. In the U.S., this can add a week or more to the wait time, depending upon the amount of freight received in a given period.

When you purchase a very large item, always make sure that the seller both packs it correctly and uses a reliable, insured carrier. A large instrument often requires special packaging and shipping arrangements as well, the price increasing by weight, up to a maximum beyond which you will need to pay full Air Freight charges, plus fees for delivery.

When buying items from any source, make certain the shipper will pack equipment in the original protective cartons. Otherwise, there is always a chance you will receive a "bucket of bolts" at the other end. In any case, pre-arrange insurance and pay for full replacement value (not so-called "fair market value," which will leave you considerably short of funds to recoup the loss).

Return Policies

Some items are of a non-problematic nature, a counterweight for an equatorial mount for instance. There is little to go wrong, and problems such as bad finish or malformation will be obvious from the first. Normal return policies usually apply, even beyond standard return periods for items that prove to have factory design defects. The regulations on such things vary from state to state and country to country.

Bottom-line considerations have made return policies a very sticky issue with most suppliers. Before buying, make sure that the shop offers some kind of a return policy of practical duration (usually two weeks) at no cost to you, or at most a small "restocking fee." Most companies offer a 30-day return policy on all items; others may impose no-return policies on certain types of goods, such as electrical or electronic items. Most outlets have managers that can make a judgment-call decision, but you will have to be very persistent to reach such individuals in a high-volume company.

Warranties

Where warranty repair is concerned, the rules are usually almost set in concrete. Never return an optical instrument to a manufacturer without first inquiring about their warranty repair routine. Most manufacturers of telescopes require a "pre-authorization" for warranty repair returns, based on a previous call or letter to the factory.

Since they generally state these policies in the Warranty or Users Manual, the factory can legally return or simply put aside any item that arrives without this code on the package. Although the manufacturer usually makes good eventually, there is usually considerable delay and a lack of sympathy for your problem when you make such an unauthorized return.

Delicate equipment easily prone to user damage (computer-pointed telescopes or CCD cameras, for instance) is a particular case in point. You generally won't know if warranty covers repair until technicians have had a look at the equipment. The process can take weeks. Mail order houses will generally not relay returned items back to the factory for out-of-town customers, requiring the customer handle all such services. You will pay full shipping at least one-way.

This is a realm where purchase through a local brick-and-mortar outlet can be of value, since many will help with returns in order to keep a customer. Some manufacturers also absorb or subsidize return costs for dealers. Moreover, most manufacturers take an authorized dealer's word that the item has a factory defect, and will repair or replace it under warranty unless the damage is egregiously user-caused, whereas it is "your word against theirs" where there is no experienced dealer to intercede.

A warranty usually applies only for the original purchaser who can prove purchase with a valid receipt, or who registered the warranty at the time of purchase and can prove their identity. Some companies require no additional documentation for returns through an authorized dealer who stocks the item in question.

Normally, resale automatically voids a standard warranty. In the U.S., so-called "gray market" optical goods from Europe and Japan are a recurring consumer problem. Usually imported by suitcase or airfreight as personal items, they are then commercially resold. Since even companies that allow transfer of a warranty require that the sale be personal, this voids most warranties.

In either case, inquiring about warranty is a good tactic when dealing with an unfamiliar source. Some companies offer warranty service only where the sales receipt originates in the country where the customer requests service. Others require only that they have authorized dealers in the country where the customer requires service.

When buying through Internet shopping or trading services, be sure to protect yourself against the rare fraudulent sale. Credible online payment exchange services offer some degree of protection when dealing by check or credit card. They will (a) verify that the holder of an account is the true customer of record at the bank they use for payment by depositing a token amount and verifying the deposit and (b) for credit purchases, verify that the card user's billing address is valid, and is the same as the ship-to address on their account record. This stops fraudulent ordering for shipment to a third party location.

CHAPTER FOUR

Used Equipment

They Don't Make Them Like They Used To

The refrain of the Old Optics Buff, "you can't buy a *good* new telescope, they just don't make 'em anymore," is a blatant misstatement. So many good new instruments are now available that it is a pleasurable occupation just to note the advancements in new models and types. Nonetheless, the acquisition of a fine used telescope can also be the prelude to years of enjoyable observing. If carefully chosen, it may meet or even exceed the performance of a new instrument.

An older instrument may also fill an observational role not easily duplicated by newer ones. Observing the solar disc with a vintage 4-inch Unitron refractor and Herschel prism stands as one example. The long-focus Fraunhofer-type objective rendered incredibly distinct and evenly graduated limb darkening, with an unwavering impression of the granulation and super-granulation of the solar photosphere. It is clear that oculars designed for the focal length and focal-plane characteristics of a particular design deliver the best performance. The old opticians used to do this. Thus, using an older instrument, its quaint oculars with glass and lens curves selected to match the hand-figured objective, is another plus one won't find in most modern telescopes. Mass production has rendered the custom ocular set obsolete, not irrelevant.

Whether you are "antique" hunting or looking for a newer instrument in the used market, exercise care and discretion. Check the back issues of astronomical journals containing equipment-test articles. These can be especially helpful when considering a used instrument, reviewed when it first came on the market. The topics below cover sourcing and acquisition of the range of types found in the market today.

It is obviously important to check the condition of a used instrument as thoroughly as possible. "New condition – only used twice!" in a sale listing makes one wonder *why* the instrument saw little use. In addition, a carefully used telescope may outperform one that has seldom been out of its case. *Normal* wear is a plus in the sense that someone was motivated to put the telescope to good use. View-home "showpieces" on the other hand, may have mechanisms frozen from disuse and optical surfaces affected by fireplace smoke or cooking vapors.

Long storage times take a toll as well. Instruments improperly packed away eventually develop damage: optics succumb to fungus growth or the outgassing of acidic contaminants in the storage container. Lubricants dry out; creeping oxidation can eventually "freeze" moving parts and bearing surfaces. An early acquisition, purchased at a sale in the Florida garage where it had been stored, had all of these problems. It was usable for a novice, but irretrievably flawed. The damages would not have occurred had the user stored it in a drier location, or at least given the glass occasional exposure to sunlight. Run through the tips under point-of-purchase testing for new scopes (Chapter 1) before committing to a used purchase.

Sources

Online trading venues have become *the* prime centers for exchange of used equipment. The combination of instant electronic communication with a broad range of amateurs interested in specific items appears to be an unbeatable equipment source. However, there are still a number of ways to acquire used equipment other than browsing the Internet. Old-fashioned talking or a little footwork at a star party can provide useful knowledge, and making the acquaintance of local enthusiasts can open the way to a smart acquisition. Other sources include:

Amateur Astronomers: Observers changing instruments or retiring from active astronomy often have equipment to sell or trade. Most amateur society newsletters carry notices by members with telescopes and accessories for sale. Major astronomy periodicals worldwide also offer personal sales listings.

Star party gatherings large and small usually feature swap tables and dealer venues, on at least one day of the event. These can be a wonderful source, and the ability to try the equipment out on the spot makes it one of the best ways to acquire a good used scope. It is also a fine way to preserve value on both sides of the transaction. One caveat: you shouldn't expect "fantastic bargains" here. Sellers tend to be experienced observers who know what they have, and expect fair value from people looking for just the right thing.

Astronomical Equipment Shops: Local astronomy shops generally buy or take used items in trade. They are a good bet for finding an item or two. Giving them a short "want list" and checking back frequently may turn up what you are looking for. Cautions: Realize you are paying something to the intermediary for the consignment. It's also more difficult to try out equipment purchased this way, since there may be an "as is" no-return policy on such goods.

Auction: Unlike most auction markets, where values can fluctuate wildly in response to collecting trends, the prices realized from instrument auction sales strike a good balance between practical value and prestige pricing. Internet auction buying is the most obvious source, with the two or three main operators handling thousands of optical items every month. The larger traditional auction houses such as Christie and Sotheby (both U.S. and U.K.) periodically hold scientific instrument sales that include some vintage astronomical telescopes and sports optics. In the U.S., Yeier Optics of Candor, New York, sponsors a series of yearly auctions that have yielded generally good reports over the years.

Classic artisanship and historic interest aside, the prices realized for these fine instruments and accessories are often well within the range one might expect for a brand new item of similar size and value. Telescopes by 19th and early 20th century makers of course are at a premium, but often have excellent optical quality and hold their worth. The potential justifies waiting for opportunities, and makes obtaining equipment at auction both educational and practical.

Caution: Auction buying is not for everyone. Experienced bidders study the online or catalog descriptions carefully, attend the preview of an on-site auction if possible, and obtain clear images of expensive items before committing. Whether buying or selling, be sure to know the Internet operator or auction house's physical location, shipping and commission arrangements, and obtain *written* copies of the contract. When selling, obtain confirmation of your reserve-bid figure.

Military Surplus Outlets: Old-timers fondly remember the surplus bargains available in the 1950s and 1960s. Giant battleship and "flak" binoculars, sighting telescopes, assorted lenses, optical mirrors and wide-field eyepieces flooded the market. However, most such items are no longer bargains. One surplus company recently offered a common type of Korean Conflict-era military "elbow" telescope for sale. These hefty 8×50 units carried an even heavier "nostalgia" price, far above their usefulness to the observer.

In fact, with the exception of specific models of military binocular (excellent naval 6×30 and 7×50 glasses turn up), most military surplus items offered at collector's prices are unsuitable for serious astronomical or sporting use. Light throughput is often poor due to age clouding and/or multiple uncoated lenses and prisms. The military makes telescopes primarily for low-power daytime or twilight use, with heavy cast mountings designed to take hard shocks. Objectives match built-in erecting systems and usually have permanent eyepieces. Despite the occasional find, most of these perform poorly when remounted for use with newer accessories. Additionally, rare-earth glass in a few vintage wide-field military eyepieces and objectives harbors potentially harmful radioactivity, only detectable by special equipment. Other than the addition of antireflection coatings, the above considerations hold true for some pseudo-military imports that are actually commercial productions cobbled together from stored or second-quality parts.

Newspapers: Don't overlook the "For Sale" columns of local newspapers, although most telescopes listed will be of the department store variety. People with no interest in astronomy sometimes acquire good telescopes through inheritance or other circumstance, and go to the papers to sell. Buyers for these items are few, especially in smaller towns. Such sellers often have little or no concept of condition or completeness of the instrument, however. If you like it, be prepared to explain condition as a bargaining point (and also to hunt up any damaged or missing parts and accessories).

Schools: Due to the increasing demand for bigger optics, electronic drives and accessories, the science departments of secondary schools and colleges occasionally dispose of old equipment to defray the expense of new instruments that serve updated needs. They are a potential source of older telescopes, particularly refractors and long-focus Newtonian reflectors of moderate aperture. Low-key inquiries may turn up an unexpected opportunity. Be cognizant that indifferent students have often subjected such items to heavy use over the years; expect to do some refurbishing and parts replacement. Don't be surprised if you also encounter considerable administrative red tape before concluding the deal.

Antiques Venues: The flea markets and antique shows held in most cities on a regular basis are a potential, but sporadic, source of optical items. The occasional useful piece may justify the time spent looking over acres of glassware, tools and bric-a-brac. With luck, one may sometimes acquire unusual or useful optics.

Caution: Be prepared to drive a hard bargain. Regardless of working condition (which, in general, they haven't the experience to judge), dealers in these venues usually overvalue old instruments based on apparent age or decorative value, but not always. A friend picked up a nice folded 80-mm refractor for $60 when he happened to glance behind the counter of an antique shop. The proprietor hadn't been able to figure out "which way to look through the darn thing."

If you are considering paying a collector's premium for an ostensible antique or vintage instrument from any source, you have a right to be particular. This should go beyond determining that the piece is in good condition. Homemade and after-market accessories of lesser quality, for instance, may have weathered to look original. Try to obtain photographs or copies of advertisements or catalog illustrations to determine the original appearance of the model and its included accessories. This research should include determining the correct style of the maker's name or logo for the period, and the standard locations of serial numbers or other indicia. In any case, use common sense in looking over the instrument. Metric-size fittings on parts of a U.S. made Alvan Clark refractor, or U.S. Patent numbers appearing on parts included with an English Cooke, a French Bardou, or a German Zeiss for instance, speak for themselves as probable additions. In the case of a piece represented as very rare and valuable, always seek the second opinion of an expert before paying the asking price.

Condition Checks and Evaluation

It is obviously important to check the condition of a used instrument as thoroughly as possible. As a first step in checking a potential purchase, run through the tips under *Point-of-Purchase Testing* (Chapter 1).

Before traveling to look at a used telescope for sale, inquire if the owner has the tools that came with it. If not, find out what sort of fittings it has, and assemble the tools you will need. In the simplest case, have a small flashlight, flat and cross-point screwdrivers, a set of hex or "Allen" keys, and an adjustable wrench or spanner on hand. If a heavy mounting has an electric drive, inquire if power is available. If necessary, bring along your own extension cord long enough to reach a power source, or a portable battery and current inverter unit with the requisite voltage and ampere output.

Also, take along several eyepieces of the type you normally use, since you will want to check the image at various magnifications with familiar oculars. Note that many older telescopes have eyepiece holders of a proprietary diameter. These can be the Japanese 0.965-inch or various inch- or metric-based standards, even thread-in styles of various types. Inquire ahead so you will know whether adapters will be required to use your oculars for testing.

Locating missing parts or accessories may result in significant unanticipated costs. Request that the owner fully assemble the mount and telescope before your arrival if possible. Items are commonly missing from a used outfit that may have changed hands before, and it is very easy to overlook such while browsing through a box of parts. Make sure the unit is complete with the optical tube and mounting hardware, mount head (with counterweight shaft and counterweights, if an equatorial), tripod or pier and its finder, if that was a stock item. Always ask if the owner's manual is available, since any special use and maintenance procedures are important to know.

Refracting Telescopes – Optical Tubes

There is slight potential for damage to a refracting optical assembly in normal use. Its cell and dewcap or glare shield generally protect the objective lens from incidental damage. The most weather resistant glass surface faces outward in most designs. Objectives stand up well to atmospheric conditions and humidity that can destroy the delicate mirror surfaces and fragile secondary mirror mountings of reflectors. These factors all contribute to a long operating life. Good telescope mounts are also made to last. The materials are usually corrosion resistant, or if not, at least well primed and painted. Optical tube assemblies are generally quite robust. Moving and adjustable parts have traditionally been made of brass, rust resistant alloys or plated metals. Good manufacturers generally fabricate such parts to close tolerances, and protect them with a durable lubricant.

A practical first step is to closely examine the physical condition of the entire optical tube assembly under good lighting. If possible, first remove any dewcap or light shield over the objective. Most are threaded or otherwise easily removable. Note the construction of the objective cell. Is it metal (usually aluminum, brass in older scopes), plastic, or a composite material? Plastic, especially if it has obvious mold-marks and an uneven finish, is characteristic of low-range manufacture – but not necessarily a reason to reject the scope if the price is right.

Other than basic construction, does everything "look" right? Tool marks, loose screws, scratches and scrapes, or paint drips on the objective cell reveal sloppy manufacture or clumsy maintenance attempts. Examine the instrument with a careful eye. Then, run through the basic checks under New Instruments (Chapter 1).

Examining for Objective Type and Condition The illustration of Basic Objective Profiles (see Chapter 1) shows the lens arrangements of several classic refracting objective types. There are three typical configurations; air-spaced, cemented and oil-spaced or "oil immersion." In the first type, three small, equally spaced rectangular metal foil shims may be visible at the perimeter of the lens. These hold the elements slightly apart, usually by about 0.003 inch (0.1 mm) in a small standard Fraunhofer-style doublet, creating the air space. The air space can be large in the case of older objectives, such as types made by Alvan Clark & Sons, some of which have the "flint" lens placed in the front position in the cell, as in the designs by C.A. Steinheil. Otherwise, it is generally less than 1mm, even in larger, more modern achromats. A thin spacer ring may also separate air-spaced lenses. This is usually not immediately detectable, covered by the cell's outer retaining lip or fitting.

Air-spaced Lenses: A mystique of quality attaches to these, since air spacing is a requisite when using different curves for the two inside-facing surfaces, so the adjacent lenses don't contact sharply at the edges or center point. Varying the space facilitates corrections for both coma and spherical aberration, yielding an aplanatic system under the Abbé criterion. However, in apertures up to about 3 inches (80 mm), both cemented and air-spaced objectives can perform equally well. Cemented types have potentially higher light throughput and less chance of internal "ghost" images, because there are only two air-to-glass surfaces, versus four in air-spaced designs.

To help in examining the optics, find a bright artificial overhead light source (**not** the Sun!). To check the number of elements in the objective, turn the upper surface of the lens toward the light source. Closing one eye, tilt the tube slightly while looking down into the objective from a short distance away. Each lens face will create its own reflection of the light, forming a "stack" of bright images. The surface curvatures of the lenses determine how the different reflections will look. Generally, divide this number by half (adding one first if it's an odd number, since the backside of the first lens may reflect very diffusely) to determine the number of lenses in the objective. A standard two-element achromatic objective will yield three reflections, one upright, and two inverted. If uncoated lens elements are separated by an air space, all three reflections will be nearly equal in brightness.

The reflections should maintain a straight lineup as you turn and move the objective relative to the light. If they don't, there is an alignment problem in the making or mounting of one or both of the lens elements. Because shipping shock and vibration can cause this type of misalignment, keep an open mind and check other examples before rejecting the model as inferior.

In cemented objectives, the optician shapes the interior-facing lens surfaces to matching curves and bonds the lenses together for edging when they are finished. Formerly, they used a thin layer of a rosin preparation or Canada balsam (purified pine resin). Balsam is still useful in restoration work (see Chapter 13 for a method), but clear synthetic methacrylate or other synthetic adhesives cured under ultraviolet or visible light have generally replaced it in manufacturing. The center reflection from this cemented boundary will usually be much dimmer than the other two.

Check cement condition by looking through the glass at a lighted wall, without an eyepiece or diagonal in the focuser. The cement between the lenses should appear clean and clear, with no yellowing, crackling, bubbles, or an appearance of colored "Newton's rings." Any of these show degradation through aging, parting, or cleavage of the cement from the glass surface. Air-spaced objectives will show the reflection pattern noted in Chapter 1.

The difference in the linear coefficients of expansion between the glass types limits the practical size of such cemented objectives. Changes under sharp temperature variation, such as transport from room temperature to freezing conditions, can cause distortion, shear the cement between the lenses, or even crack the glass itself. Unless it is one of the rare cemented-triplet Apochromats, an objective with cemented lenses should not exceed around 60-mm aperture, 70 to 80 mm at most. Some good finders of this diameter have cemented doublets of short focal length, not designed for high-power use.

A triplet lens of any type is a sophisticated design that may show five or six reflections. One doesn't commonly meet with these in a selection of optical instruments aimed at the beginner. Oil-spaced or "oil immersion" doublet objectives are another special case, and generally of very high quality. The lens boundary is filled with a microscopically thin layer of synthetic oil or silicone-based liquid. This allows slippage for adjustment to temperature variation without distortion or strain. One might suppose that having a liquid between lenses would lead to accidental leakage. The oil layer is so thin, however, that surface tension generally keeps it from migrating, and the optician seals the periphery of the lens group as a further safeguard. The three reflections from an oil-spaced doublet will be similar to those from a cemented objective, but the middle reflection will generally be even more diffuse.

Checking Baffling While holding the front of the telescope toward the light, look for an aperture stop (a black disc with a round central hole) mounted directly behind the objective lens. This old dodge blocks the rays from the peripheral edge of a poorly made objective, while giving the false impression that the observer is utilizing the entire aperture. Although Galileo was the first to use aperture stops (in front of his lenses) out of experimental necessity, there is no modern excuse for this deceptive practice: Avoid the instrument.

On the other hand, unless the tube is specially designed, true annular baffles or "glare stops" finished in flat black are necessary for top performance in standard achromatic designs, and are especially present in most high-quality designs as well, although there are exceptions. Properly dimensioned annular baffles spaced along the tube interior increase image contrast. They trap scattered light and control stray off-axis rays. Additional baffles or matte-finished threading are often mounted or machined inside the focuser tube as well, generally a mark of quality construction. Chapter 13 describes a method of checking and baffling an existing refractor tube for better performance.

Check for lens coatings: Compare the brilliant reflection from a bare glass surface, like a clear-glass light bulb or wine glass, with the subdued reflection from a coated lens to see the difference. The light you don't see reflected is passing through the lens, to your eye.

The first hints that a coating on lenses improved performance come from the 19th century. As H. Dennis Taylor wrote in 1891: "It may seem a somewhat startling statement to make, but nevertheless it is a fact, that certain flint glasses that we have experimented with by local tarnishing have been found to transmit actually more light where tarnished..."[1]

Various forms were tried until developments of the 1930s that led to modern methods. Antireflective material is generally deposited on the lens as vapor discharged in a vacuum chamber. Coatings are *very* thin, only about 120 nanometers, or roughly the thickness of 1/4-wavelength of visual light. The standard coating material for decades has been magnesium fluoride (MgF_2), which gives a violet tint to the reflection.

More recently, multiple coatings have been applied, to increase light transmission or to block particular wavelengths. Few objectives 6 inches and under are left uncoated by the manufacturer. The same goes for eyepieces. Multicoating doesn't necessarily mean *good* coating; improper choice of materials and method can actually reduce the efficiency of a multicoated lens over that of a properly single-coated ocular.[2]

You can determine the number of coated lens surfaces in an objective by counting the number of violet and magenta tinted reflections returned. Makers in previous years sometimes coated only the outer or *first surface*, of the front element. The best objectives have all air-to-glass surfaces coated. Examine the first surface coating for spotting, an oily "rainbow" appearance, or the uneven texture that indicates slipshod quality.

Objective Condition First, run through the checks for baffling, coatings, etc. outlined above and under new purchase (Chapter 6).

Carefully check the surface of the lens under a light for smoothness and intact coatings. Repeated overenthusiastic cleaning, even with the right materials, can remove patches of coating or create a haze of fine scratches or sleeks unnoticeable under casual scrutiny. These faint markings degrade the image by scattering light before it enters the lens system. Slight wear of this type will not necessarily compromise the image, but it is a definite factor in value. If the lens surface is slick, check the perimeter of the objective for small conchoidal (seashell-shaped) fractures or chips. These indicate damage from side impacts or over-tightening in the cell. If something like this crops up, be especially attentive to collimation and alignment, as described below. A tiny edge-chip can be touched up with flat

[1] Taylor – see the Bibliography.

[2] This is another opportunity to check manufacturer claims and hyperbole against reality. A good visual test is to simply line up and look down at the tops of a selection of oculars in a shaded area outside. You may be surprised at which ones show noticeable external and internal reflections, a sign that their coatings are not very efficient, no matter how many layers are applied!

black, and will not noticeably harm the image. If between the lenses, however, it may be difficult or impossible to do.

Dust and Fungus Contamination Check for excessive dust and especially for fungus growth on or between the elements, by removing the eyepiece holder and looking straight through the tube at a bright surface (**not** the Sun or its reflection!). Etching from thready fungus growth will show as a network of thin, connected lines across the circle of vision. In fact, a bad case will be immediately evident when looking into the objective from the front.

The "crown" (outer element of most achromats) is prone to this, usually first on the protected inward-facing surface (i.e. the air-space), confirmed by the experience of those who own Fraunhofer-objective refractors in semitropical climates. Fungus also occurs in air-spaced triplet objectives and can attack the rear-facing air/glass surface of oiled triplets in their tubes as well. Not surprisingly, the contagion occurs on both coated and uncoated surfaces. Whether antireflection coated or uncoated, fungus can easily affect the inner or fourth, "flint" surface of cemented objectives as well; also the interior elements of complex telephoto camera lenses. Fungus-dissolved coating forms the thready pattern, while the glass surface itself seems unaffected when examined with a magnifier.[3]

The air space is where the fungal organism receives the least ultraviolet, even when exposed to sunlight from both ends of the tube, due to the strong cutoff of rays below the near ultraviolet by the optical glass.

Within five years or so of bringing equipment into the semitropics, threadlike fungus contagion had appeared on the inside-facing crown surface of a "giant" air-spaced 80-mm finder, and on the coated, inner (flint) surfaces of three binocular objectives. Two local telescopes acquired there had the incipient problem as well. In the present author's experience in the semitropics, where heavy nighttime dew forms year-round and relative humidity can remain in the upper 90th percentile for weeks at a time, the problem attacks all glass types, including the multicoated surfaces of modern oculars and camera lenses left exposed to damp.

Life under extreme conditions of humidity confirmed the efficacy of the "sunlight fix" (see Chapter 6). It works – whether by combination of drying and the limited penetration of ultraviolet light is not certain – but no optic ever developed a fungus problem while using the routine regularly. However, where fungus is already present, nothing is possible at the user level except removal, cleaning and sterilization of the elements to halt further damage. If the glass performs well

---

[3] This has been the case since the late 19th century. H. D. Taylor, in his 1891 classic *The Adjustment and Testing of Telescope Objectives*, (cited earlier) found the problem chiefly on his [Chance-Pilkington] crown glasses. Taylor was working with uncoated lenses. He puts the cause down to the presence of soluble metallic minerals in the glass composition, combined with the tendency of the crown glass surfaces to more easily acquire and retain a film of condensed moisture, as a nutrition factor for the fungus organism (see the Bibliography).

enough to consider purchase, such a condition is cause for a major reduction in price. Professional refiguring to restore the surface is, of course, a logical option in the case of a valuable glass.

Cement Defects Opticians often cement the lenses in a smaller doublet objective. For instruments of 60-mm aperture or less this is not necessarily the sign of a bad glass. Most good refractors of 80-mm (3.1-inch) aperture or larger, however, have air-spaced objectives.

Cement Layer Defects: Lens cement poses long-term problems, due to yellowing, crazing, crackling, partial separation, etc. Examine the objective closely both in reflected and transmitted light to detect any bonding anomalies. In general, avoid objectives with cement-layer defects. Separation and re-cementing of antique objectives is possible, and optical artisans have developed methods to salvage such glasses. Given the wide range of adhesives used by various makers, however, there is no guarantee of success without damage. Still, it is a valuable exercise; Chapter 13, Making Finder and Eyepiece Cross-Hair Reticles, describes one technique.

Collimation and Adjustment – General Information

If the light converging through the objective to the eyepiece doesn't pass down the center of the tube to meet the eyepiece at its center, we refer to the telescope as "out-of-square," or decollimated. This compromises the image, and optical testing will be inconclusive. Realize, however, that the objective being out-of-square with the tailpiece is a common problem in refracting telescopes, even new ones. It's not difficult to determine whether a refractor is significantly out of adjustment, but it is a multipart problem.

First, look for a means of adjustment at the front of the lens cell. Many glasses, some by top-line makers, are sealed and non-adjustable. These require factory servicing if the objective gets out-of-square. On the other hand, adjustable cells have three obvious and equally spaced pairs of screws or "push-pull" bolts running through the rim. These should be easily accessible with the dewcap removed. If there are no adjustment fittings, as is the case with many modern glasses of 4-inch (100 mm) and smaller aperture, you're facing reduced options for correcting the problem. Still, check for means of correction. Decollimation is definitely a bargaining point in a "used" sale, or cause for factory return if new. Chapter10 describes a practical collimation field-check for refractors.

Tip: Unless the seller gives you a clear return option if a problem such as decollimation cannot be corrected, don't acquire the telescope. Moreover, visual testing of any kind will be inconclusive with a decollimated refractor. In order to check collimation, use the small flashlight and a card, and run through a simple reflection check as described in Chapter 10. If the objective is out of alignment, go back to the push-pull bolts or other means of adjustment at the front of the cell, and attempt on the spot to grossly align the glass. This will ensure that the action of the adjustment fittings is effective, not frozen by corrosion.In the worst cases, deformation of the tube or mechanical shock may have thrown the cell out of square, and you may reach the end of the adjustment threads before the objective approaches a collimated condition. If out of collimation, and there are no adjustment fittings, your only option after purchase may be to attempt adjustment by squaring off the tube, adjusting the cell in the tube with shims, or a complete tube replacement.

Lens Alignment Problems

Check the physical alignment and optical centering of the objective cell using the reflection test outlined previously. For another, more critical lens alignment test, place a small flashlight (or candle, as was formerly done) in front of the open eyepiece holder. Look into the objective at the successive emerging reflections. The bright images should cover each other concentrically when viewed straight on, and shift to form an absolutely straight line when viewed off-center in any direction. If they don't, one or more of the elements has shifted or tilted in the cell, is made of grossly inhomogeneous glass, is thicker on one side ("wedge"), or has an offset center of curvature due to improper grinding or edging. This can be double-checked from the front, with the light pointing into the objective. There will usually be a similar appearance, though not as markedly differentiated. Taking up slack in the lens retaining rings, readjusting or replacing the foil shims, or replacing the spacer ring between lenses can sometimes correct tilted elements. In the case of a badly machined cell, the strategic placement of a thin brass or aluminum shim between the edge of the tilted optic and its machined seat may rectify the problem. Properly a job for an optician, this should obviously not be attempted in the field!A professional maker usually discovers actual lens defects before a mounted glass leaves the factory. Such a compromised objective will never perform near its potential. Reconditioning requires replacement or re-figuring of the element by an expert or the factory. If you are considering purchase, a scrupulous owner may agree have the objective checked by an optician to determine what is causing the problem.

Reflecting Instruments – Optical Tubes

Mirror Condition As with lensed instruments, the optical tube assembly is the prime consideration. Both mechanically and optically, the reflecting telescope is a delicate proposition compared to the refractor. Defects, however, are more readily apparent on physical inspection, since all optical surfaces are exposed and easier to directly examine.

Ideally, an astronomical mirror blank is tolerably homogeneous and ground smooth on the back before it is figured, without thicker spots or a "wedged" profile. Any of these contribute to surface distortion under temperature change.

If inhomogeneity is extreme in a thin mirror, it may even contribute to astigmatic images by "pancaking" along an axis of weakness when the tube is at certain attitudes.

Newtonians and other compound reflectors without correcting plates are easily affected by atmospheric contamination and infalling debris. Foreign material wafting down the tube constantly settles on the mirror any time the tube is left open. This is unavoidable during observation, since air constantly flows past the mirror in most designs – the direction of flow depending on temperature and other factors. Truss-tube designs (without shrouds) avoid this flow, but are even more vulnerable to anything blowing in the wind or dropping from the sky.[4]

Inspection: The appearance of a mirror under casual daylight inspection, even one with an old coating full of pinholes, may initially be that of a bright, shiny, coherent surface. The overall "effect" is bright and the brain is often temporarily fooled; the eye doesn't inspect the actual surface when viewing from a distance, instead it views the reflected *image* of the surface, smoothing it, much as it smoothes the "pixilated" view seen on a computer monitor or on a digital movie screen. A well-coated but *badly figured* mirror also looks fine, of course, since the deviations from smoothness are virtually undetectable under normal scrutiny. Frankly, just looking down the tube at most mirrors reveals little other than the presence of excessive surface dust, or obvious coating and mechanical surface damage.

Flashlight test: Observers often shine a bright flashlight (electric torch) into a scope tube at night to make a critical examination for dew or mirror contamination. Of course, this causes even the tiniest specks of dust on a mirror or corrector plate to leap out with alarming clarity. One manufacturer advises against this in their descriptive literature, and for several good reasons. Practically speaking, most mirrors immediately attract a number of such specks, even after the most rigorous cleaning. The appearance, while annoying, won't noticeably compromise performance, if at all. For one thing, the dust only blocks light from the amount of surface area it actually covers, so the effect is of undetectable dimming until the dust motes increase to form a dusty coating.

Moreover, even a good, solidly coated mirror may give a false impression under a flashlight test when it is back-illuminated. The process will of course reveal any gross defects in the coating, also pinholes and thin spots. Variations inside the glass disc, however, will also stand out in grand contrast. This is deceptive, since most mirror discs, especially those made of Pyrex, have denser areas and striae from the manufacturing process that generally don't compromise performance.

In short, simple visual inspection may alter one's opinion, even of a supremely figured optic in need of a simple recoating job. Such direct visual methods are of little value in determining the most important factor in performance, i.e. the *surface figure*. Observers and opticians employ such methods as the star test, discussed later, to evaluate this important factor.

---

[4] See notes on airflow in a Newtonian tube in Chapter 13.

Recoating: Coating or re-coating even a small telescope mirror is a technical feat often under-appreciated by the consumer. Maintaining the coatings on large professional mirrors can be a budget-challenging feat of engineering.[5] While most amateurs will never need to ship a 2.5-meter mirror, the facilities serving the amateur trade are still rather sophisticated, although the variety of equipment may be obsolescent by some professional standards. Nonetheless, the same result is desirable: an even, well-adhered reflective coating, without pinholes and thin areas. In fact, most coaters will redo a mirror at no charge if the customer finds the result unsatisfactory.

Be aware that repeated recoating can eventually affect a mirror's surface. This is due to micro-roughness caused by particularly strong alkaline compounds sometimes used to remove the old coatings. Although other factors usually swamp the effect of these small-amplitude defects, roughness that exceeds the level of "micro-ripple" can cause measurable light scatter and degrade the image. There has been experimental documentation of such surface conditions, most recently using phase-contrast imaging after long-term exposure to strong solutions of sodium hydroxide.[6]

Optical Tube Assemblies

"Tube" is a misleading word, since mirror enclosures come in a variety of shapes. Whether it is a truss, a conventional tube, or something in-between, one first looks for solid construction and the lack of obvious damage. The optical tubes of standard equatorially mounted Newtonians are usually constructed out of one of four materials: Rolled steel or aluminum tubing, fiberglass composite, or laminated-fiber tubing made for the construction trade.

Finish color: Most tubes are finished in white for good reason; titanium oxide pigment – in gloss enamel in particular – resists solar heating even better than reflective metal finishes. At the end of a sunny day, put one hand on the roof of a dark-finished automobile and the other on a white one parked next to it and you will get the picture. The titanium white coated tube will experience significantly shorter cool-down time after sunset as well, but it's a tossup for purposes of stargazing. Many amateurs aren't convinced this is critical, and the hour or so gain in cool-down time for a large tube may not make up for the loss of pleasure in having your favorite color on your scope.

Laminated fiber: The most commonly used brand of laminated tubing is Sonotube®, although there are other makes. The material is weather-resistant, and

---

5 * A public NOAO document states: "Transportation costs increase by an order of magnitude once the size of the mirrors, in their shipping crates, exceeds the size that can be shipped in a standard container (~2.5 m). The cost (per square meter) of glass mirror blanks and the cost of the mirror handling equipment and the coating facility increase above this range. And if alternative materials such as silicon carbide or beryllium are required it may be difficult or impossible to obtain satisfactory blanks larger than 1.5 to 2 meters." (from: *Optical Design and Fabrication for an Extremely Large Telescope*, NOAO Optics Working Group, 2001).

6 Herbert Highstone, *Best of* Amateur Telescope Making Journal, vol. 2, p. 320 – see the Bibliography.

has good thermal and vibration-damping qualities. Glorified cardboard it may be, but it serves well for years if it has been finished well with a sealer and waterproof finish. Epoxy enamel over a primer is particularly effective. Renew the surface finish in any spots that have worn through to the bare fiber, a common effect in transport or on areas that receive a lot of handling. The spiral-formed layers may de-laminate if repeatedly soaked from the ends; end-sealing, waterproof end-rings and caps for the tube-ends are critical.

Fiberglass Tubes: So-called fiberglass tubes are often really a thin fiberglass resin coating over a laminated-fiber substrate, not a solid tube made of impregnated fiberglass cloth. Both styles are subject to checking and splitting of the resin surface after years of use, and heat or ultraviolet from sunlight accelerates the process. This is a point to watch for, although minor cracking or surface checking is chiefly a cosmetic issue. This is a case where spot filling, sanding and recoating with a tough, exterior-grade white enamel such as alkyd or epoxy with an infrared-reflecting titanium ($Ti0_2$) pigment is helpful.

Metal Tubes: Rolled-and-crimped steel tubes have been touted as stronger and more reliable than aluminum of similar construction, but this is a bit misleading. For one thing, steel is simply cheaper for the manufacturer. In the gauges used for commercial reflector tubes, steel is heavier and thus not used over about 10 inches (25.4 cm) of aperture. It is nearly as quick to dent as aluminum, and harder to "pop" a dent out of when it happens. Moreover, like all ferrous metals, it is prone to rust when paint or other coatings scratch or wear away. Use fine steel wool with naval jelly, followed by a good rinse, on any rusty spots. Recoat with a spot of primer, and follow with matching touchup.

End rings: A good tube of any kind should have protective rings fitted over both ends. The rings provide dimensional stability, while preventing dew saturation and edge chipping of hygroscopic tube materials. Usually these are made of milled aluminum. More recently, plastics have come into use. Missing rings indicate that the tube is either badly designed or poorly maintained.

Tube-Mounting Rings: These are found mainly on equatorially mounted Newtonians, and come in many styles. Smaller equatorials will have the flip-over padded type with a knurled nut on a threaded stud that fits in a notch to secure the tube down. These bolt, in turn, to the mounting head. To rotate the tube to convenient observing angles, just keep them loosely snugged.

Squaring up: Chapter 9 covers a method for squaring up the tube on mount cradle rings. It works equally well for all optical tubes mounted on German equatorials, including refractors, Newtonians, catadioptrics and compound systems.

Catadioptric and Compound Optical Tubes

Mirror Condition With closed-tube "cat" scopes (SCTs, Maksutovs, Mak-Newts, Schmidt-Newtonians), the primaries lurk at the base of a tube enclosed by a correcting lens. The user has a more difficult time getting at these protected optics to make a critical examination. The same holds true to a lesser extent for the Classical Cassegrain and other open-tube variants, since the mirror is a long way from the tube opening and the spider arrangements that hold the secondary mirrors. Spot checks include close inspection for coating irregularities. You will usually need to do the "flashlight" routine, bearing in mind its drawbacks.

Coatings and Mirror Materials: Types of coatings and "coating groups" applied to cat scope mirrors and correctors are legion, and the list grows as time goes on. Makers often label the tube with proprietary names for these coating groups, and the curious may consult factory specs in the manual for details. Enhanced coatings can increase the total reflectance of the mirror surface from the nominal ~90% range of standard protected aluminum, to as high as 96 or even 98%. The gain in performance is notable.

Tip: When buying a used scope, check both primary and secondary mirrors of cats produced in the mid-1980s for degradation of some early multicoating versions, which encompasses graying, spotting and even peeling. These can be rectified by recoating at the factory – please don't try removal of an SCT primary at home, unless you have experience and a fully outfitted shop with a clean room.

SCT Mirror chips: When examining a used instrument, make a check of the primary for flakes or chips at the top and bottom edges, indicating a past accident. Many SCTs have rear cells drilled and tapped for the installation of accessories. Some of the holes go all the way through. Using too long a bolt – to attach a camera mount for instance – can put the bolt right in the path of the mirror's edge at close focus. Chips or digs nearer the mirror center, or even on the secondary mirror or its baffle, are not at all unknown, so take a close look. All it takes, for instance, is clumsily dropping a 10-mm Plössl down the baffle tube with the visual back removed. Stuff happens. We won't mention names.

If you wish to remove the corrector plate (or have it done at the factory), mirror damage of this type can be flat-blacked with the tip of a hobby brush. It usually has no significant optical effect, although it devalues the instrument.[7] A bright chip, however, will throw rays across the image, and deep digs call for mirror replacement.

Corrector Plates Some disregard it as a mere "dust window." The corrector plate, however, unlike a flat optical window that has no power, is the critical element in a catadioptric system. The outer surfaces of both SCT (toroidal) and Maksutov (spherical) correctors have significant optical power.

One facet of optical tuning at the factory level involves rotating the plate (and secondary) to cancel small variations in the mirror set to give the best image. For this reason, the working optician marks the corrector and tube during assembly. The orientation of these marks isn't self evident, however, which is really too bad. Several cases come to mind, where a previous unaware user has removed a corrector for cleaning, and replaced it devil-may-care. Unless you are prepared for hours of reiterative rotation under star-test conditions (see Collimation), factory return is the best option for diagnostics of this kind. In any case, if you ever

---

7 The loss of light from the non-reflective area is the only result if the area is properly light-deadened by touch-up.

remove the corrector from a "cat" scope, clearly mark the edge and the tube so the orientation can be re-established upon reassembly.

Makers have used various types of coatings, especially on SCT corrector plates. Earlier versions were uncoated or used simple magnesium fluoride (MgF_2) antireflection coatings. Multicoatings soon followed. Like special mirror coatings, these are generally noted in the model specifications or on labels placed on the optical tube itself. An antireflection coating on the corrector plate interior is especially important in an SCT, to avoid back-reflections, which can gray out the image.

Surface condition: This is a prime indicator of the scope's previous treatment. Other than built-up sleeks or scratches from enthusiastic cleaning, indications of a bad surface include an effect of *iridescence* when viewed at an angle, like the colors of the Newton's rings seen on oily water: these can indicate freezing damage to coatings from improper canned-air use, or over-cleaning that has revealed patches of underlying coating layers. Don't confuse such damage from a merely oily surface, which can mimic the effect. Carry out light cleaning with alcohol/detergent Solution B or a good non-silicone cleaner such as those produced by Edmund, Orion, or Zeiss, and recheck the plate.

CHAPTER FIVE

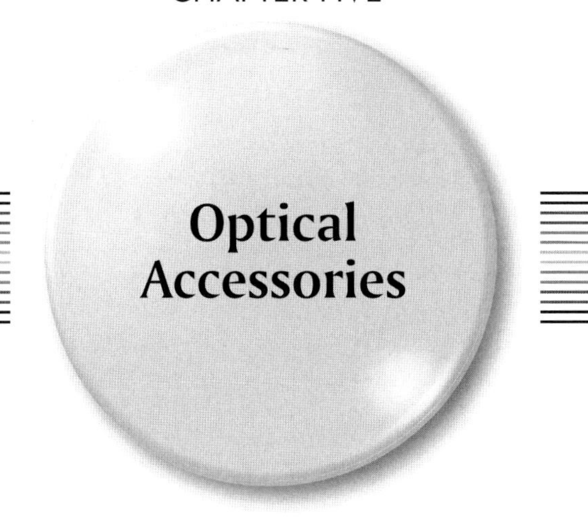

Optical Accessories

Even more than with telescopes, the selection of accessories in the market today is, to use a '60s phrase, "mind-blowing." This section is intended to give the observer a good handle on some styles and types of accessories available, but it cannot be complete. Just in the realm of eyepieces, there are dozens of manufacturers competing. The range of prices is much wider than ever, and individual oculars can cost as much as a telescope.

Eyepieces and Barlow Lenses

The basic types of eyepieces for sale today have been around for a least a century. These include the hoary 18th century Huygenian and Ramsden designs, the German designs of the 19th century: the Kellner, Plössl, Erfle and Abbé-type Orthoscopic eyepiece, as well as designs by König and Steinheil.

Several types, like the wide-field eyepiece designed by H.V. Erfle (1884–1923) and modern variations on the designs of G.S. Plössl (1794–1868) reached a state of high development for military use in the 20th century. Simplified for stock selections, or elaborated with newer glasses, they retain their designer's names as general *types* of eyepiece. The nomenclature serves to distinguish the general arrangement of lens groups, rather than strict adherence to design curves and spacing. For instance, almost any ocular that has two achromats with the convex faces nearly touching receives the name "Plössl," although it may bear only a passing resemblance to the original design by the Viennese optician.

All of the modern types illustrated were once available in amateur-class 0.965-inch, 1.25-inch and 2-inch barrels (metric-standard 24.5Ø, 31.8Ø and 50.8 mmØ,

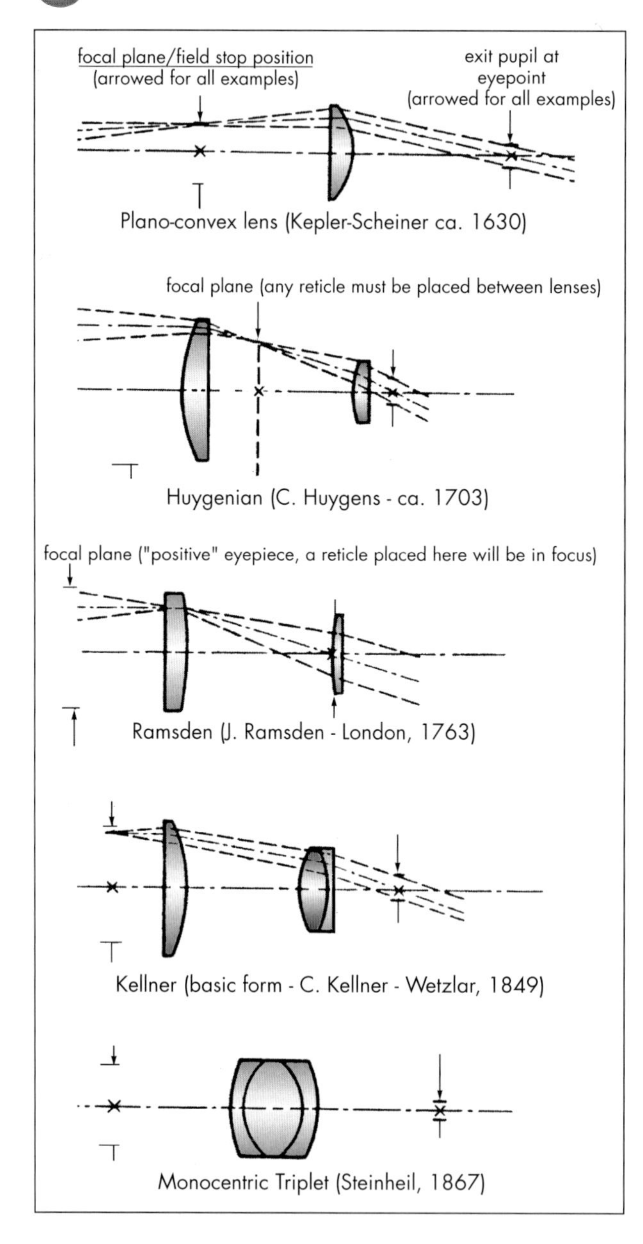

Figure 5.1. Early eyepiece types.

respectively). Eyepiece barrel diameters have been increasing in size over the past half-century, and the 1.25-inch size has now almost wholly superseded the 0.965-inch post-World-War II standard. Even the traditional Carl-Zeiss Jena optical factory had switched over to a 1.25-inch standard before its amateur astronomical equipment division closed in 1995. Some of the best classical eyepiece designs were produced in the smaller format over the years. As a result, high-quality

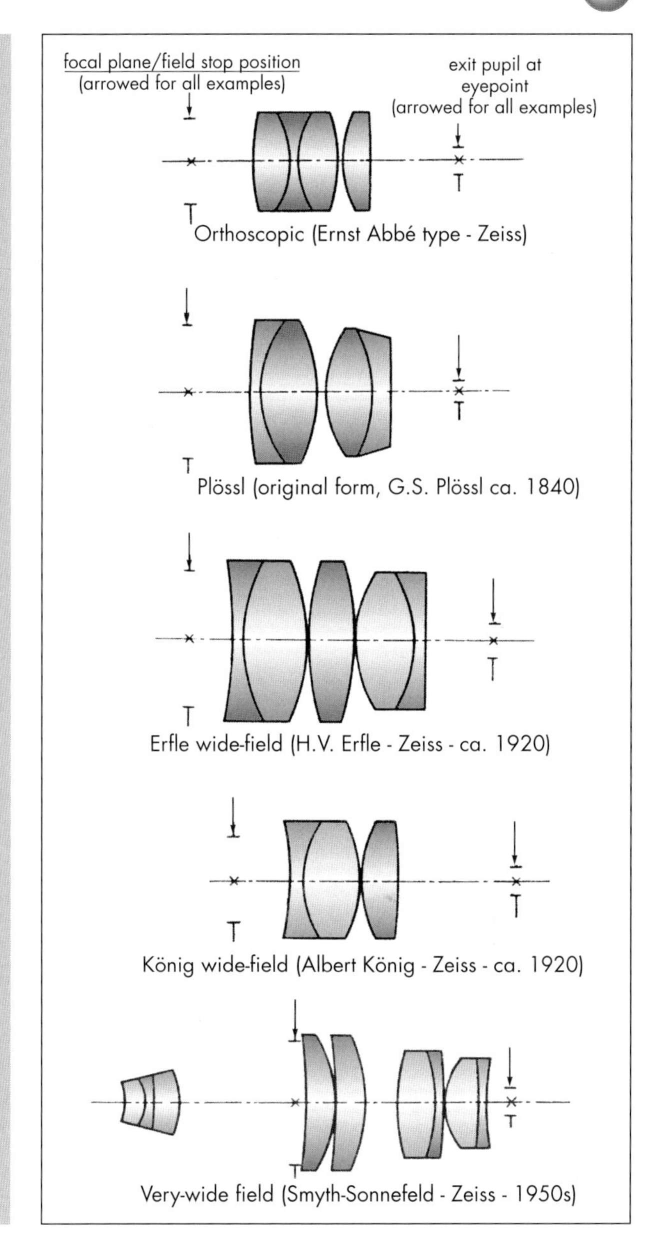

Figure 5.2. Modern eyepiece types (original forms, generally now modified for current use).

0.965-inch equipment may become something of a collector's item, once it outwears its current stigmatization as "old fashioned."

The 2-inch size, perhaps the largest that is ergonomically compatible, is now rising to supersede the 1.25. The keen amateur is encouraged to seek equipment that will handle this standard, and many new telescopes have 2-inch capability right out of the box.

Figure 5.3. Eyepieces of the three standard barrel diameters: 0.965-inch (24.5 mm), 2-inch (50.8 mm), and 1.25-inch (31.8 mm).

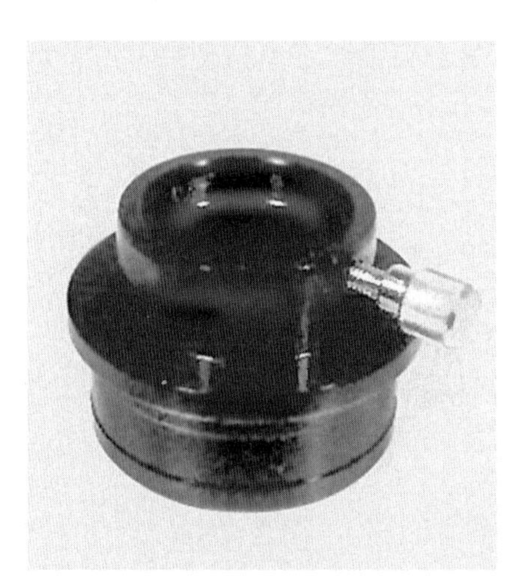

Figure 5.4. Adapters facilitate using different size oculars in the same eyepiece holder. This 2-inch to 1.25-inch (50.8 mm to 31.8 mm) unit by Tele Vue incorporates a pressure ring that grasps the accessory, centering it without marring.

Choosing Standard Oculars That Fit Your Needs

The rules for "minimum" and "maximum" magnifications are a useful guideline (although not a criterion) in choosing oculars and a practical telenegative eyepiece (Barlow lens) to complement them. We define below a suitable basic set of oculars using eyepieces of the standard Plössl type, which has an apparent field (AF) of about 50°.[1] The method is applicable to any combination of aperture and

[1] With current eyepiece designs, there are wider options for Apparent Field of course, and the matter of calculating their use within the 5×/50× parameters is a matter of simple math (see the section on *Basic Optics*).

focal length, but let's take a standard f/10 Schmidt–Cassegrain for convenience, it being one of the most popular systems in use. We may apply the 50× and 5× rules (see Basic Optical Definitions) in conjunction with the exit-pupil formula to roughly determine the maximum and minimum usable focal lengths most useful for the system. For example:

Telescope aperture: 8 inch (200 mm)
Focal length = 2000 mm (f/10)
50× Rule maximum = 400× 2000/400 = 5 mm eyepiece
5× Rule minimum = 40× 2000/40 = 50 mm eyepiece

High range: In practice, 400× is only barely usable under optimal conditions with most 8-inch compound systems. The optimal high magnification for nights of "average" seeing (and this comes from 15 years of using good 8-inch SCTs), is around ~300× or about 40× the aperture in inches (1.5× in centimeters), requiring an eyepiece focal length around 6 to 7 mm. An 11-mm eyepiece used with a 2× telenegative (Barlow) lens yields 364×, useable under excellent conditions, but without straining the capabilities of the system.

Mid Range: In this range, you might opt for an ocular similar to your eye's focal length (~17 mm); they seem particularly easy on the eye. In any case, 15 mm is near enough that figure. Doubling it with a high-quality or "premium" 2× Barlow yields a sharp view at 7.5 mm effective FL. Thus, you can cover both the mid-range and "average seeing" high range with a 15 mm. For high mid-range, the 11 mm (181×) is very useful.

Low Range: In a "basic" set, every ocular should add a useful magnification. This is a good place to add the 2-inch (50.8 mm Ø) format. On the practical side, a premium 2-inch diagonal mirror or prism will fit this type of system well, providing excellent corrections for all focal lengths *and* the ergonomic benefit of greater clearance between the observer's head and the rear cell. Now, the 5× rule-of-thumb allows an ocular of ~50 mm or about 40× for wide field use, revealing another reason for acquiring a 2-inch system. The maximum usable focal length in a standard 1.25-inch Plössl with its Apparent Field of circa 50° is about 32 mm, yielding 62.6× and a true field of about 0.8°.

You can obtain a slightly *brighter* view with a longer 1.25-inch eyepiece, but no matter; the focuser aperture restricts the true field to around 0.8°. Using a 2-inch system with a standard-field 55mm, on the other hand, gives a field of 1.4° at f/10. The field stop nearly fills the 2-inch aperture. Doubling it gives you 27.5 mm. This effective FL provides an optimal mid-range magnification for such a system – corroborated by the fact that manufacturers traditionally supply a 25 or 26 mm eyepiece of moderate quality as stock with the units. To wrap up, a suitable minimal set of oculars embracing 5× to 50× the aperture and enhanced with the 2-inch format would be:

55 mm Plössl (2-inch/50.8-mm Ø) = 36×
w/2× Barlow lens, effectively 27.5 mm = ~73×
15 mm Plössl = 133×
w/2× Barlow, effectively 7.5 mm = 266×
11 mm Plössl = 182×
w/2× Barlow, effectively 5.5 mm = 364×

This gives a ten-fold increase between lowest and highest magnifications in five nicely spaced steps, using just three oculars and a Barlow lens. One might add a

Figure 5.5. The eyepiece selection outlined in the text for the popular 8-inch (20 cm) f/10 SCT adds 2-inch (50.8 mm) format for wide-field. 55 mm, 20 mm, 15 mm, and 11 mm eyepieces, with a 2× Barlow lens, yield a tenfold magnification range, increasing in gradual steps from 1.4° at 36× to 0.14° at 364×.

20-mm Plössl (100×) to fill the gap in the mid- and mid-high range. The accompanying photo illustrates the set just chosen, including the 20. The illustration of diagonal sizes shows the 2-inch with the same 55-mm Plössl in place (Fig 5.8).

The Barlow Lens

Peter Barlow (1776–1862) was mathematics instructor at the Royal Military Academy at Woolwich. He invented and published a number of interesting

Figure 5.6. An apparent field of 82° makes the Tele Vue Nagler 16-mm Type 5 eyepiece one of the widest-field oculars yet introduced in 1.25-inch format. The design incorporates a telenegative element and has 10-mm eye relief.

optical devices, including an achromat objective that used a dialytic correcting element filled with carbon disulphide. Nasty stuff; it worked well but eventually corroded the cell. Barlow would likely be an optical footnote today but for the two-lens negative achromat or telenegative amplifier that bears his name.[2] The Barlow lens has proved to be one of the most versatile of all optical contrivances. As a negative lens, it steepens the beam from the objective, as the laser photograph illustration shows. This increases the effective focal length of the optical system at the eyepiece by a factor directly related to the Barlow's focal length and distance from the focal point of the ocular.

A Note on Eye Relief

Eye relief is the distance from the vertex of the eye's lens to the point where eyepiece forms the exit pupil, an important ergonomic factor virtually ignored for most of the last century, since it is not easy to increase using typical optical glasses without introducing image aberrations. Yet, people with enough astigmatism to require glasses while observing need eye relief of 12 mm or more to acquire the full field of view. Using spectacles at the telescope also makes it much easier for people with the common myopia/presbyopia combination of middle age to star-hop and read charts without constantly fumbling for eyeglasses.

Beginning around 1985 eyepiece designers began to add features, including built-in telenegative elements, to achieve this. More recently, adding an element of rare-earth glass of unusual refractive index has proven efficacious in both standard and wide-field oculars. Eyepieces with this quality, also called "high" or "long" eye-point, command a premium price, so one should expect to add 50% or more to acquire a selection of focal lengths as set out here.

A 5- or 6-mm Plössl or orthoscopic usually has very short eye relief, roughly equivalent to its focal length, thus, you literally push your eyelashes into the eye lens to see the full field of view. A 10- or 12-mm in a good 2× Barlow gives the same magnification with roughly double the eye relief. The telenegative component even adds a small amount of eye relief when used at its optimal focal position with standard eyepieces, since it widens the entry angle of light into the eyepiece, pushing the eye-point outward. It can also flatten the field somewhat by reversing its curvature, and help to control the off-axis astigmatism in some optical systems. Great stuff if it is well made and mounted.

Unfortunately, mid-20th century mass producers much abused the concept, and the Barlow lens lost favor for several decades. The idea was resurrected in the hands of quality makers in the 1970s. Several makers now add a third lens (the "Apochromatic" form), taking the concept further. Apparently, the lens is now around to stay. Finally available and well corrected in the 2-inch format, it forms part of the accessory selection depicted here.

---

[2] Only to settle the by-now wearisome query; I am no direct relation – my Great, great, great, great grandfather Joseph Barlow was of Midlands English stock, but emigrated to Connecticut in America, where his son John promptly turned his back on "John Bull" to enlist under George Washington and serve at White Plains.

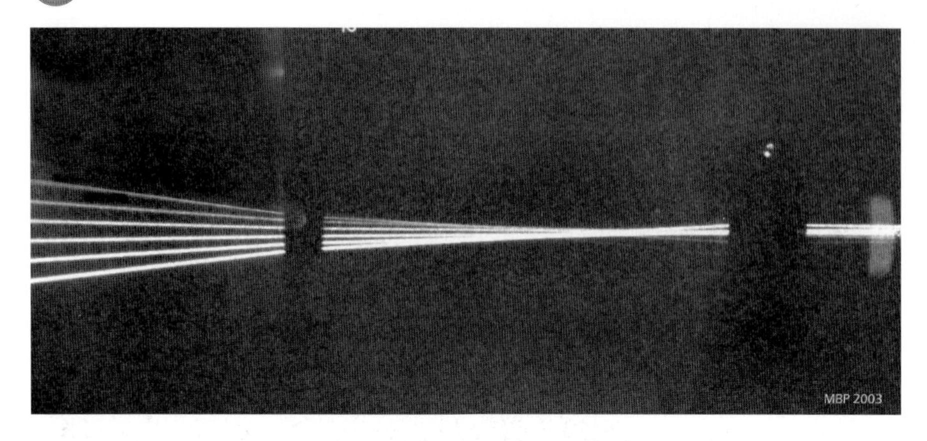

Figure 5.7. This laser tank image shows the change in angle of the parallel light as it passes through a telenegative amplifier (Barlow lens) in a Keplerian astronomical telescope system. The focal length is effectively double that of the objective used alone.

Star Diagonals and Prisms

These are among the most useful accessories ever developed for a telescope that has its focal plane centered at the lower end of the optical tube. The simple expedient of turning the light beam by 90° allows comfortable observation from a wide range of tube positions, and avoids the necessity of crouching, or even lying down, as many old-time observers did when observing near the zenith. One of the results of their universal adoption, however, has been referred to already; that is, insufficient out-travel in focuser tubes.[3] Moreover, a diagonal with problems can be more of a curse than a cure. Get the best performance out of your scope – see the Tips under *Collimation and Adjustment Techniques* for a simple diagonal-check procedure. Adjust or repair if possible, and if not simply replace a unit that compromises your telescope's performance.

Although a nicely coated prism gave better total light-throughput in the days before enhanced reflective coatings (and is easier to clean and maintain) mirrors now seem to have gathered the majority of devotees. This was not always the situation, and at least one major manufacturer switched at one time to supplying prism diagonals as stock accessories, due to the dearth of good, cost-effective diagonal flats. Suppliers now seem to have a better handle on mass-producing better flat mirrors without spherical distortion. Ultra-high-efficiency coatings reach >98% light transmission in some cases. The mix between mirror and prism diagonals is about equal; even the lowest-price units may include either one, and research reveals no convincing or lasting consensus on whether a prism or mirror gives better performance.

---

[3] See Drawtubes, under *Optical Tube Features*, Chapter 8.

Figure 5.8. The three standard sizes of star diagonal 2.0-in. (50.8 mm), 1.25-in. (31.8 mm), and .965-in. (24.5 mm). Eyepieces mounted to demonstrate scale.

The illustration shows the size differential between the three common system diameters. All the formats provide excellent viewing when high-quality components are properly selected.

Finders

The finding telescope or "finder" has a wide, low-power field of view, and usually fits onto the optical tube assembly. The user aligns it with the main optics to visually acquire objects for closer observation through the main tube. Larger models can also function as wide-field observing instruments or camera (astrographic) lenses.

Dioptric Finder Scopes A good, small refracting finding telescope or finder is the prime optical tube accessory. Yet, underpowered hard-to-focus

Figure 5.9. 50 mm finder–component type offers variable power and use of additional standard accessories.

finders with non-interchangeable components remain the stock standard, even for some rather pricey instruments. These will get you up and running with a scope, but are barely adequate for basic starhopping and centering. Visual observers who want to learn the sky deserve a better, brighter view, if only to make visual location of the area surrounding an object easier. To improve the situation, look to aftermarket suppliers.

True "component" finders, like the vintage 50 mm from Lumicon illustrated here, were once a rarity. Recently, other types of have appeared on the market with 1.25-inch (31.8 mmØ) attachment systems that accommodate standard oculars, including illuminated-reticle types. This makes precise centering of objects in the finder possible under dark skies where non-illuminated crosshairs fail.

Alternately, a standalone spotter unit of f/5 or f/6 can be adapted. The type is readily available at comparatively low cost, and works well for low power, wide-field scanning. Rings or blocks to mount them may require a bit of lookup, but they are available, or can be made with castoff components from the swap table.

Unit-Power Finders Elongated rectangular gray boxes started appearing on the tubes of scopes here and there on the observing field in the 1980's. Wandering over after dusk one evening and looking along the tube of a 10-inch Dob, I got a first glimpse of the small red target in the sky that heralded a breakthrough in telescope pointing methods. Californian Steve Kufeld had really hit on something in 1979 when he decided that the aircraft "head-up" target display (HUD) he had acquired from a surplus outlet was too heavy for his telescope. Apparently, Kufeld wasn't the first outside the military to adapt the concept for pointing, but his home-manufactured Telrad® was the first such unit marketed with the amateur observer in mind.

The "unit power" or 1× finder has now become a staple of observing, with at least a dozen designs on the market.[4] Some use a bright "dot" projected at infinity, others use their own proprietary reticle designs; most have brightness adjustable to suit conditions. People become so adept at using them that many seldom or never use a magnifying finder at all; they simply slew the dot or target to the chosen area by starhopping, then use a low-power eyepiece in the telescope to center the object.

Hints: Keep extra batteries on hand, as it is very easy to forget that the unit is on when set at low brightness. Models that incorporate flat projection screens usually have screen shields available, useful when temperatures reach the dew point.

Visual Filters

Color Band-Pass Filters

The telescope eyepiece filter is an accessory of long standing, with examples of "smoked" and colored bottle glasses used to dim or alter the appearance of the image as far back as the 17th century. The item of commerce now comes in several standard eyepiece sizes. The most basic filter, the Neutral Density or "ND Filter" is often included with telescopes as a Moon Filter. These work much better than "smoked" glass, which I tried once by holding a microscope slide over a paraffin candle to deposit soot on it. The slide darkened the lunar glare when placed before the eyepiece, but the view was not particularly clear!

Modern color filters derive from photo technology developed before dye-layer color film. Photo processes like Carbon, Tricolor and Carbro stacked layers of light-sensitized dyes or pigments on paper or glass to form a realistic color image. The photographer printed these from black-and-white positives or negatives of the same subject taken through various color filters. Certain filter hues eventually proved most useful in translating natural colors into dye or pigment.

Wratten-Standard Color Filters: Astronomers soon realized the enhancing effects of these color filters. Since there were no standard astronomical sizes, observers adapted various photographic types for their visual work, usually gelatin-and-glass "sandwich" filters. Today, the most widely used series of color filters (there are several alternate series' by Schott and others) takes its name from the English inventor and photographic pioneer Frederick C. L. Wratten (1840–1926). Kodak purchased Wratten's company around 1910 and later applied his name to their standard color filters.

Some companies began making standard "drop-in" filters designed for camera equipment.[5] In the 1970s makers began supplying a number of the astronomically useful Wratten colors in cells that would fit into Japanese-standard and 1.25-inch (31.8 mm) eyepiece-barrels. Mass-dyed filter glass – chiefly produced by the Hoya

---

[4] The author edited the testing of a number of available models – see *Sky & Telescope*, June 1996, "A Look at Seven Unit Power Finders" by David N. Regen.

[5] These still fit the rear cell recess of the C-90, a small Maksutov produced by Celestron.

glassworks in Japan and Schott Glaswerke of Germany – soon made the older glass-gelatin color filters obsolescent. Nearly all eyepieces now contain threads to accept them, although thread-pitch variation between manufacturers is still an aggravation.

Observing guides and manufacturers often recommend certain colors to enhance features, such as variations in the methane-rich layers on Jupiter, or on the reflections of substances such as the ejecta rays on the lunar surface. Beyond such general recommendations, filter choice is a matter of individual physiology. I have known people with such strong color sensitivity that they cannot abide color filters at all; they continue to be conscious of the hue, to the exclusion of contrast detail. To give another example, my color vision is weaker than average in the 550 nanometer green range that most people find brightest, with a stronger response in the near-ultraviolet range. The Wratten 80A Blue, often sold as the "all-around" filter, is nearly useless, since my vision already enhances contrast in that region, while green and orange filters are quite useful. Appendix B, Practical Formulas, contains a short list of the Wratten series, with representative data.

Nebular and Other Dichroic Filters

A *dichroic* filter is a non-polarizing absorption or interference filter, generally made by vacuum deposition of multiple thin films of mineral or metallic vapor on optical glass, using the same type of equipment used for antireflection and reflective coatings.

Professional facilities such as Palomar and Lick Observatories adopted the earliest forms of "nebular" dichroic filters as a stopgap against increasing light pollution that was fouling photographic exposures. The filters "darken" the sky by selectively reflecting or "cutting out" luminance at the strongest wavelengths produced by common municipal and home-security lighting sources such as mercury vapor and high- and low-pressure sodium lamps, as well as part of the natural background oxygen glow of the sky.

Industry sources relate that Dr. Jack Marling of Lumicon in California first made broadband versions of these multilayer interference-coated nebular filters available to the amateur market in eyepiece and camera filter sizes. Also called Light Pollution Reduction filters (LPRs), typical units have transmission efficiency peaking between 85–95% at a wavelength swath centered on 5000 Å in the blue-green region. This area is unaffected by the strongest pollution sources, yet encompasses the wavelengths of strong emissions from oxygen (O III at 4959 and 5007 Å) emitted by common emission nebulae such as Messier 42, the Great Nebula in Orion. The firm eventually made specialized varieties available, narrowly tuned for oxygen III, which is especially useful for detecting and observing planetary nebulae and bright objects such as the Lagoon and Swan in the Milky Way. They also produced the hydrogen beta (H-β, 4861 Å) emission line filter, useful on a few objects such as the Horse Head in Orion and the California Nebula in Perseus.

Several firms soon began supplying their own broadband varieties, each having slightly different widths of bandpass, and different visual characteristics. Most

Figure 5.10. Dichroic filters: a continuously Variable Color Filter and three eyepiece filters by Sirius Optics. Contrast Enhancement and Minus Violet filters also made by the firm represent a new development in thin-film applications for amateur astronomy. Courtesy: Sirius Optics.

producers also began making make the high-contrast "narrowband" styles, that darken the sky even more by homing in specifically on the oxygen III and hydrogen β emissions. While working as equipment test editor at Sky Publishing, it seemed high time someone did a scientific comparison of the models. Editor Dennis di Cicco arranged a meeting with John Guerra of Polaroid's Optical Engineering Laboratory. Guerra and I worked with his colleague Ed Bunker to produce accurate spectrophotometer transmission curves for the models provided. The results correlated well with parallel field-testing.[6] These filters greatly extend the viability of backyard and back garden deep-sky astronomy and for detection of the fainter nebulae, even under darker skies.

Sirius Optics in the Seattle, Washington area has recently produced a series of new-type dichroic eyepiece filters for amateur astronomy, one of the few notable innovations along these lines in recent years. Among these is a Minus-Violet (MV) filter that reduces the secondary spectrum effects of short-focus achromatic objectives. They produce the MV filter coating on various substrates, including a Neodymium-doped glass that cools the color rendition for a natural balance, and on refractor objective elements for OEM use. Baader Planetarium produces a similar filter to proprietary specifications in their facilities in Germany. Sirius has also applied dichroic technology to the CE (Contrast Enhancement) filter type which enhances the primary visual wavelengths, and make a continuously variable dichroic color filter slider that has garnered good journal reviews.

Color dichroic filters: These are useful for certain visual and photographic applications, notably for color separation in color CCD work, since they have higher light throughput than equivalent glass types that pass the same bands.

---

[6] Presented along with field test results in Nebula Filters for Light Polluted Skies, Philip Harrington, *Sky & Telescope*, June 1995, p. 38.

Performance Note: Wavelength and total transmission through thin-film interference filters varies measurably with the angle of incoming light. Transmission and wavelength remains stable for a given instrument, but is slightly different between instruments of different f/ratios, or when telenegative or telecompressor units are used with the same instrument, due to the change in the entrance angle of the light cone. This change is practically demonstrated by the different views through identical nebular filters when used alternately in fast and slow optical systems. In general, tight narrowband filters are more sensitive to slight changes in angle, and perform relatively better at higher f/ratios (f/8 and above) where the incoming light beam is closer to normal incidence (90°) to the filter surface.

Solar Filters for White Light

Decreasing solar radiation to safe viewing levels, while preserving resolution, has occupied the efforts of amateur solar specialists for a century or more. Before the

Figure 5.11. Two "white light" solar filter types. Left: Baader Planetarium's AstroSolar plastic film is mounted in a homemade 100-mm cell on a small refractor, with a sample of the material. The finder is provided with its own small AstroSolar filter, as an aid to acquiring the solar disc. Right: A Carl-Zeiss Jena SFO-80 photo-visual mounted glass filter, in a case with Neutral Density filters. Both types give a view of the Sun in natural color.

development of thin-film coatings, the best systems available admitted the Sun's full radiance through the scope then filtered it near the focal plane. Often, reflecting surfaces placed at specific angles such as Brewster's induced polarization and dimming.[7]

The once popular Herschel-prism arrangement, for instance, images the reflection from the bare glass surface of a narrow-angle prism, requiring strong neutral density filtration before the eyepiece, since the wedge passes about 5% of the solar radiation, and also needs heat-convection or absorption arrangements to handle the reflected infrared.[8] For this reason, observers used small telescopes almost exclusively. Statistical sunspot counts were (and are, for consistency), done through small apertures such as the 80-mm achromat used by Wolf in Switzerland.

"Mirror sunglasses" for the telescope: Solar filters made with vacuum coating technology based on advancing the notion of the "half-silvered" mirror arrived in the mid-20th century. If glass could be coated to divert 50% of light, as in a mirror beamsplitter, then thicker films of reflective metal could obviously exclude more, down to the level needed to protect the eye from solar radiation. Various metals (mainly stainless steel or chromium) deposited as a suitably even thin-film on glass could be precisely calculated to reduce transmission to the neighborhood of 0.001% (1/100,000) and less, equivalent to a Neutral Density factor of 5.0 on the established scale. The glass filters developed for photographic use range lower, around Neutral Density 3.5 to 4.0, and require additional ND filtration for safe visual use.

As good as it was to have something available that didn't heat up the inside of your telescope, resolution problems lurked. After all, a glass solar filter is an optical window placed at the aperture. It must be of sufficient homogeneity, clarity and flatness to pass an undistorted cylinder of light to the objective. Diffraction-limited front-aperture filters, such as those supplied by Questar Corporation since the 1950s, or the Zeiss SFO unit illustrated here, are made to strict tolerances and costly to produce. Again, with optical glass, you pay for what you get. Manufacturers persisted with conventional substrates, achieving better surfaces in a much more economical form. Still, there was that aggravating problem of flatness, since this is the hardest surface for an optician to produce.

We have known for decades that extremely *thin* material doesn't diffract light to a noticeable degree, even when it isn't particularly flat. Very few materials that are only a few mils thick have the dimensional stability to carry a thin-film coating, however. Thus, the solar filter made of flexible film was a long time coming. When it did, it came in the form of coated polyester, which has high dimensional stability. It was found that as long as the surface itself was smoothly

---

[7] For water (refractive index of 1.333), glass (refractive index of 1.515), and diamond (refractive index of 2.417) the critical (Brewster) angles are 53, 57, and 67.5 degrees, respectively. Found by $n = \sin(q_i)/\sin(q_r) = \sin(q_i)/\sin(q_{90-i}) = \tan(q_i)$ where n is the refractive index of the medium from which the light is reflected, q(i) is the angle of incidence, and q(r) is the angle of refraction.

[8] Discussion and comparison by Richard A. Buchroeder of Baader, Unitron, and vintage Alvan Clark & Sons Herschel Wedge diagonals appears in *Amateur Telescope Making Journal*, #17–18, p. 52.

Figure 5.12. The author observes the low sun with the Zeiss SFO white-light filter mounted on a Zeiss APQ 100/640 Apochromat. For safety reasons, all auxiliary optics such as finders should be covered or filtered.

coated to the required ND factor and not stressed, even a wrinkled polyester film of Dupont Mylar® would pass a safe, barely aberrated wavefront. Inexpensive solar filters made of coated polyester film, such as the Tuthill Solar-Skreen™ filters, pioneered in using this substrate. This maker developed safer methods of using it, doubling layers up to put the coatings on the inside. Another breakthrough came with the development of Baader Planetarium's AstroSolar™ film (see illustration). Unlike standard polyesters, which add a bluish cast, it yields diffraction-limited transmission and a naturally colored view of the Sun.[9]

Although an expertly fabricated Herschel or other reflection-based system may reveal a slight improvement over any aperture-mounted filter, including higher contrast and detail in features such as solar granulation, few observers use them any longer. The potential eye hazards and heat buildup in the optical system have led most manufacturers to drop them from their lines in favor of over-the-aperture alternatives.

Cleaning Solar Filters Glass filters: Cleaning and handing of thin-film coated materials is subject to variation depending upon the manufacturer. Most glass solar filters have an optical glass outer surface either uncoated or with standard antireflection coatings, cleanable with the methods one would use for any lens.

Do not immerse glass solar filters in cleaning solution, or clean the interior faces of glass solar filters in any way other than by air jet. The coated interiors are generally delicate, exposed, thin-films deposited without overcoating.

---

[9] See *A New Standard in Solar Filters*, by Alan MacRobert, *Sky & Telescope*, September 2000, p. 63, for a comparison of plastic-film-based filters.

Don't risk eye damage. If you even suspect coating damage from cleaning, carefully check the filter for holes by holding it well away from the face, and inspecting for pinholes by passing it between the unaided eye and the solar disc. A few pinhole breaks in the coating can be touched up, as recommended by certain manufacturers, with a spot of black enamel paint or similar opaquing medium. If a number of such tiny, bright pinholes or larger defects appear, don't use it – return the filter to the manufacturer for inspection and possible replacement.

Solar Filter film: Clean filter film only as recommended by the supplier or manufacturer. Although damp cleaning, followed by careful inspection for gaps and pinholes has been recommended by some sources, light dusting with air jet is the only completely safe method. Follow with a close inspection for coating breaks.

Note on Narrowband Solar Filters Solar observers use the term "white light" to mean relatively full-spectrum light, versus narrowband observations. Amateur narrowband observation is usually done in red light at the hydrogen-α wavelength, or in the violet light of Calcium for specialized applications. Beyond simply viewing the Sun for aesthetic reasons, narrowband solar observation is a highly technical field with its own literature.

Other than gentle surface cleaning of the exposed elements (usually optical windows) as recommended by the manufacturer, refer maintenance and adjustment of all narrowband solar equipment of this type to the factory or experienced technicians. A good compendium covering many scientific aspects of the work is *Solar Astronomy Handbook* by Beck, et al. (see the Bibliography).

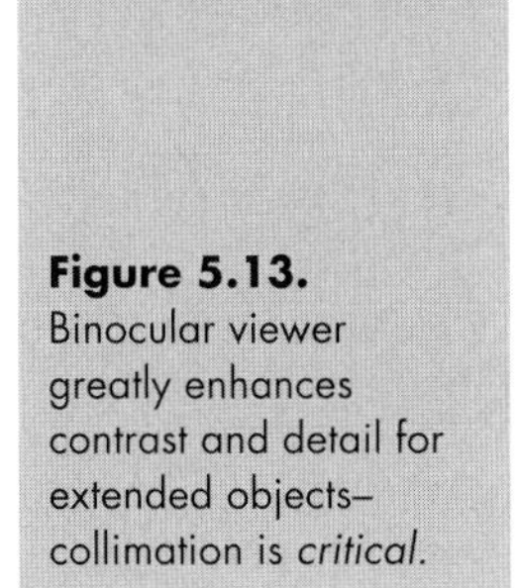

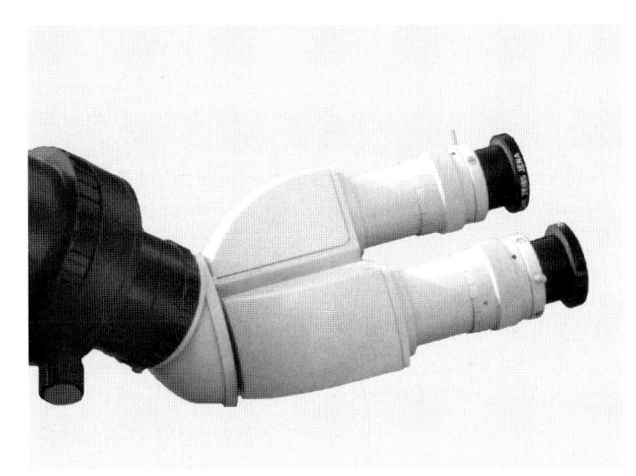

Figure 5.13. Binocular viewer greatly enhances contrast and detail for extended objects– collimation is *critical*.

Specialized Visual Accessories

Telecompressors

Originally developed in order to create "faster" systems for photographic use, telecompressors or "focal reducers" decrease the effective focal length and focal ratio of an optical system, with a consequent widening of the true field and brightness attainable for imaging. Most are for use in Schmidt–Cassegrains, other compound telescopes and refractors. Unlike the Criterion telecompressor unit illustrated, many of them also correct the focal plane by flattening it for imaging use, and are properly termed "reducer-correctors."

While telecompressors and reducer-correctors will theoretically work for optical purposes in most telescopes, they require sufficient focuser in-travel to access a point ahead of the focal plane, where the telecompressor is designed to fit into the light cone. Thus, standard units are not generally useful with stock Newtonians, Mak-Newts, or other systems with a limited focusing range.

A purely mechanical drawback is the housings, typically threaded to match a manufacturer's proprietary standard and requiring adapters for use in other company's systems. An exception is standard SCTs where the rear-cell threading is compatible.

Binocular Viewers

The binocular viewer is one of those "luxury" accessories that well repay the investment with enhanced viewing comfort and utility. Once used, the "bino" quickly becomes an indispensable addition to the observer's arsenal. When well collimated and adjusted for personal ergonomics, binocular viewing is restful to the eye, and physiologically enhances the perception of detail by engaging both halves of the optical nerve/brain system. Binocular attachments are especially useful for lunar and planetary studies. They retain the full resolution of the aperture, and the reduced brightness of the image due to splitting the principal beam is not a significant factor.

Figure 5.14. A thread-in telecompressor (focal reducer), designed for the rear cell of a Schmidt–Cassegrain telescope.

Many units have a built-in or attachable 1.5× or 2× compound telenegative lens (Barlow lens) to increase effective focal length. The steeper light cone reduces the effects of slight collimation errors, adapts the focal point to accommodate the additional air block introduced by the beam-splitting system, flattens the field overall, and provides high magnification using longer eyepieces with good eye-relief. Naturally, one must also figure the additional investment for matching oculars into the cost.

Eyepiece Micrometers and Analog Measurement

The first significant instruments to enhance the practical scientific use of the telescope were developments from the principle of the "telescopic sight" invented by the Yorkshire amateur William Gascoigne around 1634. Gascoigne, although inspired by the sharp appearance of a web spun across the focal plane of his eyepiece by a spider, developed his concept using metal indicators moved by a screw adjustment. It wasn't known until Robert Hooke's publication of it in 1655. Hooke was apparently first to employ wires, and the wire or web micrometer was born. This "filar micrometer" was the irreplaceable staple instrument of positional astronomy for centuries, but came to the end of professional use with the advent of digital imaging and measuring methods.

Since industrial spinoffs have always been the lifeblood of amateur instrument development, finely made "analog" instruments of all types are becoming more difficult to acquire as digital units supplant them in industrial applications. Micrometers such as the one illustrated have been available from various custom makers over the years, but usually for only a short period, since work of this type is now done with digital equipment and software. The Rolyn Company (see Appendix) has supplied a standard type for many years.

Figure 5.15. This vintage eyepiece micrometer by Astronetics is a "bright field" type. It uses an OSM micrometer head with a sliding non-illuminated etched-glass reticle and 0.1-mm measuring scale; the primary use is for calculating the sizes of planetary features and solar features. Micrometers with illuminated "webs" are generally used for precise double-star measurement.

Ocular Turrets

One of the niftiest inventions ever was the ocular turret, which allows mounting a selection of eyepieces and switching from one to the other at will. The concept has been around for at least a century. The device is usually based on a standard star diagonal with a mirror, 90° prism, or erecting prism of the "roof" type mounted at the hub. One acquaintance refers to the ocular turret as the "eyepiece pre-fogging apparatus," since the oculars are all exposed to dewing when mounted.

The mirror or prism is the heart of the device. As with any star diagonal, it is truly useful only when this is of the absolute highest quality. Good ones have been available from various companies over the years, including Unitron, Celestron, Vixen, Zeiss and a few small custom makers.

Figure 5.16. Zeiss 100/1000 with SFO-80 solar filter was used for this digital photo of the 3rd contact of the solar Transit of Venus through dawn haze. June 8, 2004, Rainwater Observatory on the Natchez Trace, French Camp, Mississippi: Panasonic Lumix DMC-L33 and a 40 mm Abbé Orthoscopic eyepiece. Photo: D. Martin; setup by the author.

Section II

Care and Maintenance

CHAPTER SIX

Lens Optics

Introduction: Glass as a Lens Material

Most optical instruments employ lenses in one role or another. They are the principal optical element in refracting telescopes, most spotting scopes and binoculars. The lensed finderscopes and interchangeable eyepieces used in telescopes are the workhorses of observation, subject to all the environmental and human factors that affect the entire instrument. This chapter presents a range of environmental challenges to lens performance and longevity. Practical procedures follow.

In addition to its obvious fragility, glass is a chemically sensitive material. For this reason the optical industry has established physical standards to characterize the relative resistance of different optical glasses to abrasion, scratching and breakage in addition to climatic and environmental insult. Glassmakers carry out testing on all the optical glasses they manufacture. Guidelines show that surface changes and staining of lenses from the action of water combined with soluble materials, principally acids and strongly alkaline compounds, can be quite severe.[1]

---

[1] Schott lists the basic resistance qualities under two-letter "R" codes:
"CR" – climate resistance classes 1–4
"FR" – staining resistance classes 0–5
"SR" – strong acid resistance classes 1–4, weak acid resistance classes 51–53
"AR" – alkaline resistance classes 1–4

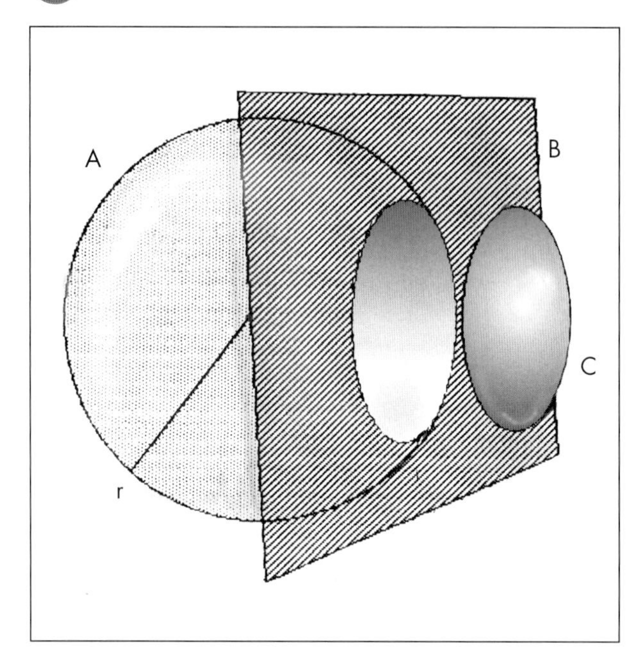

Figure 6.1. The first description of a true lens was in the optical experiment by Ibn Al-Haytham (Alhazen) as described in his 11th-century *Kitab al Manazir*. Early sources such as Vitellio were unclear about the function of this lens. **A** is a glass sphere of radius "r" cut by a plane, **B**. **C** is the resulting plano-convex lens of radius "r" that acted as a *contact magnifier* to enlarge black markings. (Clarification by A. Sabra, Harvard University)

Even by pure water damages some kinds of glass. Fortunately, only a few of the so-called "abnormal" glasses – along with crystalline compounds such as the lithium fluoride and sodium chloride (salt crystal) used in exotic optical applications – are actually water-soluble or even water-absorptive (hygroscopic).

Commercial makers of instruments as a rule use the most stable and resistant types of glass available for a particular optical design. The glass and other materials used in modern refractor and binocular objectives – as well as the types used in most eyepieces – absorb no appreciable water from the atmosphere, even over long periods. They are, however, subject to surface attack from the acidic and alkaline solutions that form in the presence of water in combination with both external pollutants, and with hydrogen-ion-soluble components of the glass itself.

Lens coatings are sometimes referred to as "protective coatings." This is a misnomer based on popular misunderstanding of advertising by certain manufacturers. Light-transmission coatings are only millionths of a centimeter thick, and do *not* significantly protect lenses from physical wear. In fact, many of the layers in a modern lens multicoating are softer than the underlying glass itself. On the other hand, light-transmission coatings help to protect the lens from some types of chemical attack.

Environmental Factors

People recognize that the severe air pollution found in urban and industrial environments can significantly damage many materials. A trio of environmental culprits: humidity, air pollution and airborne particles, can severely affect lenses. We can add the biological agents pollen and fungus to this list. The principal cause for overall concern is environmental acidity. Moisture in the nighttime air, acidified by the same sulfurous combustion emissions that produce acid rain, inevitably condenses on surfaces that cool below the dew point during observation. If this acidic collection of droplets remains on optical surfaces for long periods, their action will eventually spoil lens coatings, and may then cloud or etch the underlying glass.

Many other environmental factors have effects on equipment as well. Temperature fluctuation, aside from being a major hindrance to good observing by changing the figure of glass through thermal expansion and contraction, can also be deleterious to the equipment. Continuous temperature cycling causes stress in the metal and other materials that make up the telescope. Direct sunlight exposure eventually oxidizes and deteriorates coatings and optical surfaces as well.

Intrusion of Water and Water Vapor

Moisture can and does invade mechanical assemblies, optical tubes and lens cells, doing damage. Various dew control techniques work, as covered later on in the book, but nothing can truly protect the telescope from atmospheric humidity 24/7, as they say. Why, then, don't telescope makers just waterproof all their instruments? This is a fair question. The answer is not simple, and the subject justifies a digression for the sake of clarity. First, there is the matter of traditional construction. With refractors, most lens cells over 60 mm in diameter house two or more separate objective lens elements. These are usually held apart in the cell by foil shims or a spacer ring of metal (sometimes fiber composite or plastic), forming a narrow *air-space* of a few thousandths of an inch, generally about 0.003 inch (for a 60 to 100-mm objective).

Cells for air-spaced objectives are not usually waterproofed, an exception being those used in some top quality spotting scopes where all-weather field performance demands it. Yet, as excellent as these sports optics are, an astronomical telescope needs to be better. The typical high-quality 80-mm prismatic spotter with a zoom eyepiece yields magnifications between 15× and 60× in practice, perhaps 80× at the most. An astronomical refractor of the same aperture and relative quality is designed to preserve a near-perfect star image at more than twice this magnification; on the order of 2.2× the aperture in millimeters or greater than 160×. Given highest precision in the fabrication and surface figuring of the optical elements, imaging performance at this magnification level is further dependent on the *precise* control of sub-millimeter changes in relative placement of the objective elements. Changes in temperature that expand or contract the materials can alter this placement with deleterious effects on the image. Accordingly, the coefficient of expansion is a highly critical factor in astronomical lens and cell design.

Production cost becomes a major consideration. Providing a properly engineered and truly waterproof mounting for an astronomical objective or corrector, especially for apertures of 4 inches (100 mm) or greater, would add prohibitive cost. Even oil-spaced refracting objectives, where the lens elements are in virtual contact against a thin synthetic oil film, are water-resistant by nature but not waterproof. This is because objectives subject to the high performance requirements of astronomical use cannot be simply pressure-fitted into a tight cell using a gasket or sealants as a waterproofing medium. Such a procedure introduces dangers of warping the elements under pressure, and of deviations in the optical axes due to unequal compression of the sealing medium.

Another reason refractor and catadioptric telescope tubes and cells are not hermetically sealed is to accommodate for changes in atmospheric pressure. Anyone who has witnessed the "whoosh" of air rushing into waterproof equipment cases when they are opened after descending from altitude can imagine the potential engineering problems inherent in counteracting the effects of vacuum on the positioning, even breakage, of large glass objective elements hermetically sealed into a tube or cell.

The general solution for more than a century has been to simply let the cell "breathe." Refractor and catadioptric telescope designers add a small clearance between the inner wall of a precisely configured metal cell and the outer edges of the glass elements. This allows for expansion and contraction under temperature change. The brass and aluminum used for most refractor cells have expansion coefficients near enough to that of the so-called "normal" crown and flint glasses to allow this. Even with the slightly higher coefficient of premium lens materials such as Calcium Fluorite and Extra Low Dispersion (ED) glass, this clearance need only be slight, a few thousandths of an inch. As for watertight fit fore and aft, some definitive older texts state that the lens elements should even "rattle" a bit in the cell when shaken. I have yet to see a modern factory optical tube that fulfills this hoary requirement! In any case, lens elements are held in place using the *least* amount of axial pressure that will prevent them from shifting or tilting. Thus, moisture intrusion remains a potential problem for most refractor and catadioptric telescope owners.

Damp can invade the objective cell or optical tube assembly by two mechanisms. First, air from the exterior can slowly diffuse around the lens or corrector plate edges into the air space, where it will condense its moisture on the inner glass surfaces when the temperature drops. Additionally, droplets of condensation can join, pool at the glass/cell interface, eventually invading the space around or between the elements by capillary action. In both cases, any dissolved materials deposited when the moisture dries can begin to form a coating on the lens or corrector edge or surface. This can and does happen with eyepieces as well. The appearance under daylight inspection of iridescent interference fringes or "Newton's rings" – rainbow patterns that look like an oil slick on a pond – are an early indication of this condition between lenses. This iridescence will begin to appear when the thickness of the film approaches about $\frac{1}{4}$ the wavelength of visible light.[2] Avoid such problems by proper usage, cleaning and storage techniques.

---

[2] Don't confuse this with a similar appearance in finders, small spotters, or binoculars stored under hot or very humid conditions. Whether due to harsh storage or simple aging, the adhesive cement used to join smaller objective lenses can deteriorate and begin to pull away. This leaves a microscopic gap between patches of the cement and the glass that can give rise to the appearance of Newton's rings. Correction involves separation of the elements by heating, cleaning, and re-cementing – see the procedure in

(*continued on next page*)

Airborne Abrasives

Corrosion and abrasion from airborne salt, sand and mineral particles are particularly destructive to both optical and mechanical components. Observers near seacoast and industrial areas must be especially concerned with preventive maintenance. Natural airborne sand and alkaline mineral particles are especially troublesome to the observer in windy arid regions. Dry dust is often pervasive in the atmosphere there, even at mountain altitudes. In desert climates where condensed moisture is not a problem, optics often end the night coated with a fine layer of alkaline dust and grit. Fortunately, this is easier to deal with over the long term than wet conditions. Unless the lens or corrector experiences windstorms whipping the particles across its surface, abrasion and sleeking doesn't generally take place out in the environment; it usually happens when improper cleaning methods are applied. Safe techniques are discussed in the following chapter.

Fungus Attack

Certain fungi propagate by releasing spores that float freely on air currents. Spores inevitably land on exposed optical surfaces during observation. Additionally, in spite of sterile alcohol- or acetone-based cleaning solutions used by manufacturers, fungal organisms sometimes bloom even on seemingly clean interior lens surfaces. One strain seems to be especially adapted to growing on optical glass. Manufacturers admit that these may sometimes enter the tube or cell in contaminated lubricants or sealants used during assembly. Of course, leaving the focusing end of a refractor or "cat" scope's optical tube open to the air for long periods invites trouble of this kind.

Prolonged darkness and damp are the favorable growth conditions for this threadlike fungus that eventually forms a delicate tracery across a lens or prism. The by-products of fungal metabolism etch a permanent network of shallow channels in coatings and glass, spoiling optical performance, appearance and resale value. This is a particularly insidious problem for observers living in warm, damp climates. Intrusion of atmospheric acidity into the optical tube enhances the potential for such growth, because the organisms flourish in a slightly acidic environment. A simple prophylactic procedure to forestall such attack is covered in the section on *Lens Optics – Light Cleaning* below.

the Equipment Projects section for a possible solution. Another problem is that temperature change generates shear forces across the cement layer between lenses of different expansion coefficients undergoing temperature change. Over time, or under thermal stress, this can separate or even crack cemented elements. This is why air spacing, or oil spacing or "immersion" is preferred for larger objectives.

Human Factors

Eyepiece lenses and barrels in particular pick up perspiration, oils and acids from the skin, not to mention eye fluids and, occasionally, eye makeup. Skin oils can be quickly removed by general safe cleaning methods. Fortunately, the same considerations that make most cosmetics non-allergenic also render them neutral in acidity and thus relatively harmless to coated glass. Most forms of advanced cleaning will remove such material from lens surfaces.

Lens Optics – Light Cleaning

Dealing with Moisture and Dust Particles

Consider one Chicago AstroFest, a very humid summer night, and a weary dealer who sealed up and locked all his equipment in water resistant cases as dawn approached. Upon opening the cases at high noon for display, the interiors were so soggy that every eyepiece, accessory and optical tube had to be spread out and dried in the sunshine, literally for *hours*. Imagine the situation if this had been the last night of the meeting, and the drive home a matter of days!

In any case, allow all damp surfaces to thoroughly dry before the instrument, eyepieces, star diagonal or other accessories are capped and stored. Never just cover up and leave a moisture-laden objective, especially in warm temperatures. In this closed, humid chamber, saturated air and condensed droplets carrying pollutants and dust easily work their way into the lens cell, where intrusion between optical elements and interior dampening of the tube are nearly certain.

The sensible routine when bringing a dewy optical tube indoors is to dry it off with a towel or chamois skin. Remove the tailpiece accessories and cap the focuser tube. Let the uncovered objective dry out, pointed downwards to avoid falling dust. During cold weather, cover the objective *before* bringing it into a warm environment. Whether the surface is wet or dry, this tactic will avoid further condensation of moisture on the cold lens. When the objective has warmed to room temperature, uncap it to allow any residual moisture to evaporate.

After air-drying, carefully attempt to blow off any light dusty particles that have accumulated, without touching the lens surface. The safest method is to use a jet from an air bulb. These are available at photographic equipment outlets. Alternatively, use an "ear syringe," an inexpensive item found at most drug stores.

Air Compressors: For cleaning at home and on the field where house current (AC) is available, the author has used a small air compressor of the type sold for artist's airbrushes. Having a hose with an inline dust and moisture trap, and provided with an adjustable "brush" body or an air nozzle with a trigger, it produces safe pressurized air for optical cleaning. A spare auto tire fitted with the hose adapter available from airbrush suppliers, and provided with a moisture and dust trap, works well in the field.

Eventually the observer may notice a heavier layer of dust on the lens surface, perhaps forming patterns along the edges of dried condensation. This will quickly become dense enough to noticeably scatter light. Such a layer of dust also accelerates the process of "dewing up," as droplets condense more readily over dust grains than on a clean surface. A lens in this condition is overdue for careful cleaning.

Using Canned "Air"

The cans of compressed gas sold under various trade names can be very useful, even indispensable for certain cleaning tasks where plain, pressurized canned air is not available. Environmental laws have caused changes in the propellant used over the years. Flammability is no longer a problem with most brands; current formulations use difluoroethane or other non-flammables.

Occasional careful use has been uneventful; yet, problems remain – enough that some experts advise confining use to cleaning tasks that don't involve optical surfaces. First, the expansion of the gas as it leaves the container nozzle creates intense cold, chilling the target. The quick temperature change can stress glass, and a thin lens or mirror could conceivably crack; a safe minimum distance is 8 to 12 inches (20 to 30 cm).

A second consideration is the tendency for liquid propellant to spurt from the can if held at an angle away from vertical while spraying. The deposited liquid can leave "wire-edged" stains from impurities as it evaporates or sublimates. The effect is especially noticeable on first-surface mirrors and Schmidt–Cassegrain corrector plates. Another danger, listed by law on the labels, is the possibility of frostbite on skin accidentally sprayed with the liquefied gas. In short, use carefully on optics. Around the shop, a filtered air compressor or a squeeze-bulb cleaner will avoid problems.

Light Cleaning

If you decide your objective lens surface or corrector plate needs a light cleaning, proceed carefully. Always remove or retract the dewcap first if possible; this will make the process less awkward and the results easier to judge. Retract any extendable lens shields on astronomical binoculars or spotting scopes.

- Remove dust with careful use of a photographer's blow-bulb.
- Remove stubborn particles by careful touching (not swiping or dragging!) with a fine sable or camel hair watercolor brush. The "flags" on the natural bristles catch errant dust efficiently.
- Leave stubborn flecks for wet cleaning; they won't affect the image unless the buildup is severe.
- Frequently purge the dusting brush of loose particles by "flapping" the bristles against a clean edge.
- Wash the brush from time to time in distilled water and let it thoroughly dry, preferably in sunlight.

- Store brushes and other cleaning materials in a plastic bag or other dust-tight container.

Breath-cleaning: This method of final cleaning works well for eyepiece lenses, objective lenses, corrector plates and small mirrors. After removing all grit and loose particles first, it causes no damage to sound coatings or surfaces.[3] The optical surface should be near room temperature, as it will take and hold condensed breath for several seconds, and yet allow evaporation. Keep in mind: 1) As in dusting, use only a finest quality natural bristle (sable or soft camel hair) brush, with the lightest pressure possible. 2) The brush must be ultra-clean before starting. 3) Thoroughly clean the brush afterward and store away from contamination between uses.

First, turn the tube to view the lens or corrector surface in reflected light; carefully air-clean or dust the surface. After removing all loose particles, mist your breath onto the lens surface by "huffing" with open mouth, not by blowing through the lips, which can lead to spittle drops.

Immediately use a pressure-free, gentle pulling action with the soft, dry bristle tips to carry the condensation in short strokes across the optical surface. Dry the brush on clean lens tissue or a long-fiber sterilized cotton pad after each stroke, repeating until no more moisture is removed. Old hands say attempting to fog up the glass with breath is a good check for cleanliness, since breath-fog will evaporate in a flash from a truly clean surface at room temperatures.

Leave spots of material not removed by these simple methods (tree sap or insect spots, for instance) for the more thorough cleaning procedures below.

Warning: lens or eyeglass-cleaning solutions, especially those that advertise "antifogging" properties, often contain film-depositing silicones or soapy additives that degrade coating efficiency and may prove near impossible to remove. Tested exceptions on the current market are the solutions sold by Edmund Industrial Optics, Zeiss and Swarovski, available from the company or its dealers. Wet cleaning of optical surfaces is a painstaking job. For recommended methods, see *Advanced Optics Cleaning* in Chapter 3.

Sunlight Prophylaxis

Occasional exposure to heat radiation from off-axis sunlight will dry out optical elements that have absorbed damp, avoiding the internal fogging that can happen when the temperature drops and controlling internal fungus growth. A session every few months should be sufficient, perhaps more often under extremely humid conditions. This proactive step works for all lensed and catadioptric optical equipment subject to damp intrusion, including refractors, SCTs and other compound closed-tube telescopes, binoculars, spotting scopes, eyepieces and camera lenses. Depending on the glass types, some proportion of ultraviolet

[3] The exception is freshly aluminized or silvered reflective surfaces that have no protective over-coating. Brushing these is never a good idea, as noted by several authors, including Thompson, p. 159 – see the Bibliography.

will pass through, sterilizing the fungus organisms that are prone to grow on lenses, corrector plates and internal prisms in binoculars or spotting scopes. The procedure also helps keep the internal relative humidity below the growth threshold – most fungi will not bloom below 65–75 % relative humidity.

Preparation: Even an instrument that seems dry on the interior may have unseen moisture trapped around or in mechanical elements. When a closed, damp optical tube is heated by sunlight, moisture may redeposit as droplets during cool-down, causing unsightly patterns on the inside surfaces of all but the most scrupulously clean optical elements. To avoid problems, be sure to open the focuser tube, removing the ocular and any other blockages, such as prism diagonals or installed filters. If you know a tube has damp intrusion, leave it open (with the opening pointed downward to avoid infalling dust) for several hours beforehand and during the exposure to sunlight. In any case, do not close any instrument up until it has returned to ambient temperature after sunlight exposure.

Do not leave the optics unsupervised at any point of this procedure, as accidental shifting may carry the optical axis into the Sun's path. Do not point the telescope directly at the Sun during the procedure!

Procedure: If indoors, do this through a window, preferable open in order to allow more of the solar radiation to strike the optics. In all cases, let the instrument or component settle to ambient temperature before exposing it to the Sun. Remove any dewcap or glare shield, if possible, and firmly mount or block the component up, pointing the optics at an angle that allows the sunlight to glance at a considerable angle through the lens group. With a telescope, set the angle so that diffuse sunlight hits the tube wall just behind the objective, far ahead of focus. Also, take care to point the component above or below and behind (eastward) of the Sun's path on the Ecliptic; otherwise the Sun's apparent motion could carry it across the optical axis and focus burning rays. Limit exposure time to a few minutes, and check the surface temperature by hand during the procedure; don't allow the tube or cell to become too hot to handle.

Flooding by Water

Quick action is critical with an objective or "cat" tube that has been internally soaked by rainwater, snowfall, or water accident. This has happened to observers at major star parties, once when a flash thunderstorm came through the observing field, another time through a whirlwind followed by rain. Luckily, there were folks around with tools and calm advice.

First aid for a soaked refractor or spotter is to immediately remove the entire objective cell or corrector plate from the tube if possible, taking care for cleanliness. For smaller air-spaced objectives, the methods in Chapter 8 may be adequate, but please read the following points before taking any action:

- Set any disassembled components to dry on clean towels or absorbent cotton clothing.
- If objective or corrector plate removal is impossible, remove any dew shield, then remove the entire tailpiece, or visual back and focuser assembly, from the tube.

- *At the very least,* open up the focuser by removing any components and reducing adapters, then drain out any water present.
- Place a clean, lint-free cloth or paper, such as old toweling or clean paper towels, into the tailpiece opening to encourage osmosis and absorb water as it drains. Don't touch the backside of a refracting objective with such material.
- Finally, leave the tube assembly *open* to the air, with tail downward, to dry the tube interior

Larger instruments: Remove the objective cell if possible, shake any loose water or material off, then allow the cell to completely dry out as soon as possible, preferably in daylight (not in direct sun) or under gentle warming from a portable space heater.

Do not attempt to disassemble and clean complex lens assemblies unless you have both the confidence and experience to do the work, the proper tools and a clean, dry facility at hand. As soon as it is dry, immediately return a valuable multi-element objective to the manufacturer or send to an optical repair facility for cleaning and reassembly.

Do not ship a wet optical component: The leaching of soluble components from lubricants, cements, or fungal growth – which can bloom almost overnight – may compound the problem beyond repair before the facility can attend to it.

Cemented objectives: The owner can usually remove, dry out, clean and re-install simpler units, such as cemented finder objectives, with excellent results. Follow the directions for disassembly and cleaning of cemented objective lenses in Chapter 8.

Lens Optics – Advanced Cleaning

The advanced cleaning of lens optics, like many other procedures, takes care and forethought to produce good results. There has been a recent revival of the use of "strippable coatings" such as collodion to do the job. As this seems to be a topic of enduring interest, advanced mirror cleaning in Chapter 7 covers the procedure, which is valid for both lenses and mirrors. If one cannot remove the objective lens from its cell for cleaning, an "edge-dam" as recommended under *Collodion Cleaning* for mirrors may not prevent the mixture from creeping around the lens edges, and collodion cleaning is *not* recommended. Additionally note that, anecdotally, the efficiency of collodion for cleaning lenses is slightly less than that for mirrors, with a higher likelihood that bits of the collodion may need to be removed with a tape loop, as described for mirrors.

Cell and Objective Lens Mechanics

Someone who has acquired and maintained a new a refractor for a time should already have looked it over pretty well and may be reasonably familiar with its type and construction by the time it requires advanced cleaning. A recently acquired used instrument, however, may need detailed attention from the outset. The straight-through reader will have gained some familiarity with the basics from reading the section on Point-of-Purchase Checks. Someone who is using this book as a look-up resource should go over that section first.

The task of cleaning a standard, small two-lens objective is not a particularly daunting one. The first question to ask is what type of lens is it? Chapter 2 covers this more completely. Most modern objective lenses fall into the categories of achromatic (roughly, "two-color corrected") or Apochromatic (roughly, "three-color corrected"). Both are designed to cancel false color and other faults in the image, and exist in two, three and even four-element forms. The semi-Apochromat is an excellent design with performance falling in between the two types, generally having two or three lenses in the group. As noted earlier, these terms are only generally descriptive, and cover a broad spectrum of image quality.

When well designed, the excellent imaging characteristics of these lenses make them superior in certain respects. Like standard achromats, the Apochromats and semi-Apochromats may be air-spaced, cemented, oil-spaced, or a combination. Manufacturers generally employ "abnormal" exotic glass or minerals in their manufacture. Their cost and risk of damage by the non-expert is proportionally greater. While the recommendations for checking an objective apply to all refractors, maintenance, especially disassembly, of the newer double- and triple-lensed Apochromatic and semi-Apochromatic objectives lies beyond the general recommendations given in this book. The observer intending to acquire an Apochromat as a first telescope should rely on the manufacturer for recommendations on adjustments. Also, refer oil-immersion (oil-spaced) objectives, large air-spaced objectives, large Apochromats, or antique objectives to an experienced working optician for cleaning and refurbishing work.

Note on Procedures: Small Objectives – Disassembly and Handling in Chapter 8 covers the removal process for small air spaced units. It is important to mark the orientation of the lenses, just as before separation of a cemented lens, as illustrated in Chapter 13, Separating and Re-cementing Small Achromats.

Cleaning a Disassembled Objective

A Standard "Wet" Method Carry out objective cleaning in as dust-free an environment as possible. Obviously, this will not be possible in the field.

Tools and Materials List

Tools:

- Tools required for objective cell removal and lens removal (screwdrivers, spanner wrench, etc., see Chapter 8)
- Tweezers (for lens-shim or spacer-ring manipulation)
- Blocks padded with lint-free cloth, one to support each element while drying
- Bowls or containers of adequate size, one for each lens

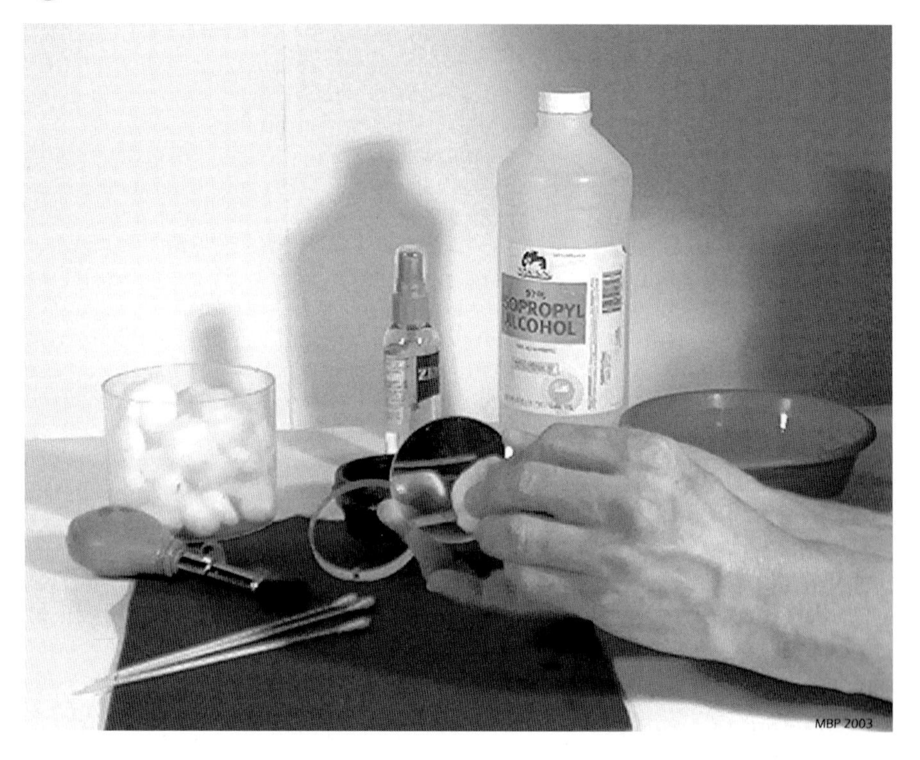

Figure 6.2. Cleaning the crown element of a small air spaced achromat, using surgical cotton and isopropyl alcohol. The selection of recommended materials shown include surgical cotton balls and swabs, an air-bulb cleaning brush, and a proprietary lens cleaning liquid.

- Covers of Plexiglas™ or clear acrylic (Perspex™ in the U.K.) to overlap the edges of each container
- A clean work surface with lint-free cloth padding
- Clean, oil- and dust-free flat camel hair or sable brush, $\frac{1}{2}$-inch (12 mm) width or larger (or air bulb with natural brush)
- Air compressor, air-bulb, or "canned air" (use precautions)

Materials:

- Surgical cotton balls or cotton in roll form, sterilized (pharmacy grade)
- Cotton swab sticks (available from optical suppliers, or wrap your own on small-diameter (1/8-inch/3-mm) dowel stock or craft sticks.
- Lint-free cloth pads for each element in bath
- Extra pads for draining, two for each element
- Standard Solution A: distilled water/detergent solution, 1 quart (liter) or sufficient to cover lenses by 2 inches in baths.
- Standard Solution B: alcohol/detergent solution, same amount

- Distilled water, 1 gallon (4 liters) (will suffice for several cleanings)
- Rubbing alcohol, isopropyl, 91% or stronger, 1 pint (0.5 liter) (will suffice for several cleanings)
- Pure grade acetone (dimethyl ketone), and eyedropper for application (optional)

Caution: Do *not* use glass or Pyrex® bowls or a hard ceramic sink for the work – the potential for fracturing a lens edge is great when working with small glass items. Ideally, use a large, flat-bottomed plastic or thin stainless metal containers, larger than the objective, one for each lens. This will allow you to soak the elements at one time without the glasses coming in contact. Choose containers that can be easily tipped and fully rinsed. Do the work near a standard sink with a tap for solution disposal.

Lens cell: This is a good time to clean the lens cell. Be sure to remove any greasy material from the threads of a thread-on cell with a toothbrush dipped in VM&P Naphtha or pure mineral spirit before placing in a detergent bath. Replace the lubricant with a scant amount of lithium bicycle-type grease before reassembly.

Smaller cells and retaining rings can be soaked in hot water and detergent, sponged to remove soil and rinsed well in tap water, followed by distilled water. Air dry, then, protect the parts in new plastic bags, after dusting with a clean brush and air jet. Food storage bags of the freezer type work well. Wash a larger cell in the same way.

Cleaning Procedure: After marking their orientation with pencil and separating the elements, check to make sure that any adhesive shims are firmly in place on the surface of the lenses. Carefully remove any loosened shims or air-spacing rings with tweezers, and set aside in a safe place in a small container. If a shim comes loose during the cleaning, pick it out of the bath with the tweezers, rinse it in distilled water and set it aside undisturbed to dry.

Place the lenses in containers pre-filled with a Solution "A" bath on a soft, lint-free cloth pad at room temperature for an hour or more, agitating at intervals. Tilt the elements up one by one to determine when surface material is loosening, using latex or rubber gloves to avoid slips.

Make a pad of surgical cotton about 1/3 the diameter of the lens, and lightly brush the surfaces underwater to remove any clinging material. Turn the lenses over on their pads and repeat. Be careful not to scrub at antireflective coatings. Small sleeks may seem like dirt: abrasion will only exacerbate any existing disturbances.

Note: Be *very* careful not to abrade the coatings present on the between-lens surfaces. Through experience (and for unknown reasons, possibly a cooler application process of the coatings for interior surfaces, whereas harder coatings may be used on the outer surfaces), these tend to be a bit softer than the outer coatings, and one may easily alter them by scrubbing. If any fungus etching is evident on the inside of the convex (usually "crown" glass) lens, cleaning will not remove it. In short, *soak, and let the solution do most of your work.*

Take the lenses out and lay them on a clean, dry portion of the cloth-padded surface. Rinse the containers well, and fill with distilled water. Hold the lenses in the distilled, agitate well to rinse for 20 seconds or so and remove, placing each on a clean, dry section of cloth. Remove the lenses and toss the distilled, rinsing to remove any cotton lint traces. Dry the containers with lint-free cloth.

Fill the containers partially with Solution B (alcohol-detergent solution) enough to cover the lenses fully, but not much extra. From this point, keep the lids on the containers to slow evaporation of the isopropanol.

Place the lenses in the Solution B baths on fresh sections of lint-free cloth. With gloves on, pick up and agitate each lens a few times, checking for remaining contamination. Examine the surfaces carefully. Using a cotton swab or two, wash lightly over any dirt particles or resinous material that remain. Don't scrub if they don't come off with a light touch.

Lightly graze over the entire surface of both sides with clean cotton balls or pads dipped in the solution, then remove the lenses after a few minutes. The elements should be thoroughly clean of most oily or water-soluble contamination by this point, but don't address any remnants now.

Pour the Solution B from the containers down the sink and rinse with plenty of tap water. Rinse the containers with distilled water, and then turn all but one upside-down to dry.

Hold each lens in turn over the remaining container, and pour a slow stream of pure Isopropyl alcohol over the surface, turn over and rinse the second surface, then tilt the lens on a clean section of cloth pad with the upper edge just touching a padded block to air-dry. If necessary, quickly pick up any trace alcohol droplets with the tip of a fresh cotton swab.

The lenses should be sparkling clean at this point except for possible slight wire-edge iridescence from drying solution. Buff lightly with clean cotton in a circular pattern from the center outwards, and examine them closely when they are dry for remaining contamination. Any remaining spots are resin, coatings, or oil not susceptible to water, alcohol, or detergent.

Keep your gloves on. If they are wet from previous solutions, wash and dry, or change them.

Acetone spot-cleaning: Do this work with adequate cross-ventilation, allow no open flames or coals in the room, and keep the acetone away from any plastics such as plastic containers, lids, or furniture finishes.

To address any remaining spots you can determine are *not* coating defects, soak a small ball of cotton in the pure acetone with the eyedropper.

Holding the lens by the edges with clean gloves and viewing the surface at an angle, lightly draw the acetone-charged cotton over the area of the spot, repeating several times, and swirling outward toward the lens edge to distribute the acetone so no wire-edges form. Do this on each surface of each lens that is not spotless, and finish by buffing lightly in a circular pattern from center to edge with a clean cotton tuft.

Remove any residual cotton lint or dust from the lens surface by light brushing followed by a jet of air, or by air-bulb brushing.

Residual fogging may be removed by breath-cleaning, as described earlier, using more fresh tufts of cotton, followed by air jet or air-bulb brushing.

Reassembly: Replace the objective in its cell under good lighting, reversing the order of disassembly. Replace the air-spacing ring. Replace any shims that have come loose at a 120° separation around the edges of the inner surface of the concave lens. Carefully bring the lenses into contact, checking for dust between them by holding the unit up to a brightly lit wall or viewing against a bright sky. Use an air jet or brushing with an air bulb until no specks remain.

Place the outer lens cell over the padded pedestal you used to remove it, then set the assembled objective on the pedestal and carefully bring the cell up to enclose the lenses. Remove from the pedestal, start the retaining ring or other retaining fitting, and snug down to a finger-tight fit. Don't hold the objective in its cell with significant pressure.

Replace the cell on the optical tube assembly, checking the inner surface carefully for any dust that may have collected in this operation, removing it with an air jet or with an air-bulb brush.

CHAPTER SEVEN

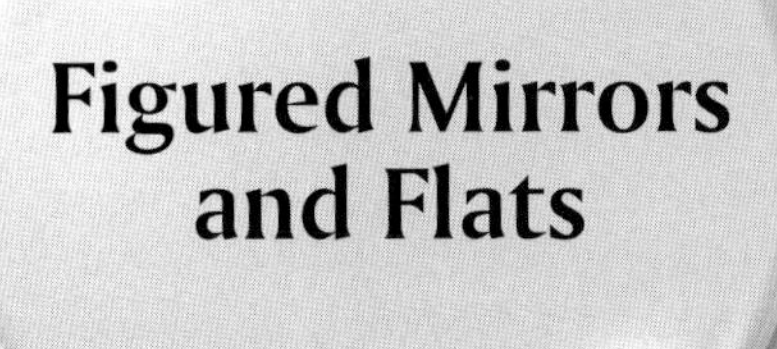

Figured Mirrors and Flats

Environmental Factors

Many observers use their closed-tube Newts for months, keeping the highly visible secondary flat clean, yet neglecting to protect and check the heart of the telescope, the primary mirror. Serious deterioration of the coatings can happen quickly in certain climates with acidic atmospheres or a heavy atmospheric load of acidic pollen or fungus spores. This can be costly when time comes to rectify it. If nothing else, the downtime while your mirror is off at the recoating facility results in lost observing opportunities. Prevention, as usual, is the best cure. Both figured and flat mirror coatings and substrates are subject to the same problems, and the solutions are the same for both.

Physical protection: Its situation at the base of the tube shields the Newtonian primary from mechanical damage. Capping the front of the tube when not in use and keeping an eyepiece or plug in the focuser would seem to provide all the protection necessary. Yet, this gives a false sense of security. The cliché "out of sight, out of mind" was never more applicable.

The facts: In addition to the constant, slow infall (or updrafting, depending on the climate and tube design) of airborne dust and contaminants through the reflector's tube, there is also the matter of nighttime formation of acidified dew on the reflective mirror coatings. Both contribute to steady surface deterioration, in spite of protective overcoating of most modern mirrors with silicon monoxide.

Dealing with Moisture and Dust Particles

Condensed droplets carrying airborne pollutants and dust are constantly drifting down inside the tube of a Newtonian reflector, during observation and whenever it is uncovered. As with lenses and corrector plates, allow all damp mirror surfaces to thoroughly dry before the instrument is stored. In fact, since the primary mirror in particular is usually less accessible than a lens objective surface, buildup that can spoil reflective coatings takes place almost unnoticeably. Dusty residue holds in moisture, in turn encouraging the blooming of fungus spores. As with lenses, warm temperatures and moist air have the effect of accelerating such incipient problems. Fungus contamination can be especially problematic along mirror edges, and in absorptive areas such as mirror-clip pads and edge supports.

Dry a closed-tube Newtonian with a towel or chamois skin before storage. After this, in secure outside environments, it may be better to simply turn the tube horizontal, leaving it partially open to dry out over the course of the morning following an observing session.

If you must bring a reflector indoors from the cold for protective reasons, clear out sufficient room in an unheated (winter) or un-cooled (summer) space to let the telescope dry thoroughly. A clean shed or garage is perfect for such purposes. When practical always cover the optical tube top and bottom if possible, *before* bringing it into any significantly warmer or colder environment. In cool weather, wait until the optics have warmed to room temperature before uncapping the tube to assist drying, This will avoid both heavy condensation of moisture on the cold mirrors, and unwanted thermal stress on components.

Unless you have complete facilities – with purified water and a good, stable work surface at hand as discussed in the following section – resist the temptation to remove and clean the mirror under such circumstances, even if it appears to have picked up considerable pollution overnight. Simply remove any auxiliary equipment and oculars, and cap the focuser. Then wipe down the tube, fittings, cradle and mounting parts thoroughly, letting the instrument stand to complete drying out.

An older, dry and well-washed towel draped across the tube opening will stop dust intrusion, while allowing the components to dry out safely. Even better is an overall linen or light canvas cover put on when the wipe-down is complete. Such precautions are especially valuable in very dusty environments, such as high desert where winds are a factor.

Fabrics such as heavy linen bed sheeting or light cotton artist's canvas (*unsized* cotton duck), make good covers. Terrycloth (Turkish toweling) would seem to be an attractive choice as well. The problem is that toweling is almost *too* absorptive. It stays wet a long time, doesn't breathe as well as standard light woven fabric, and sheds lint. Wash and thoroughly dry any fabric used for making instrument coverings several times before doing the sewing or fabric bonding. "Breathable" nylon or polyester fabrics also work fine for protection, but many types have a slightly abrasive surface that can cause surface-finish damage if blown repeatedly against the telescope surface by daytime wind gusts.

Routine checks: Carry out the following routine check procedure several times a month, especially in seasons when you are actively observing:

After air-drying, carefully examine all optical and interior light-baffling surfaces that are open to the weather. One can use the same methods of "dry" cleaning as with a lens, although a slightly different tack must be taken with closed tubes, since the mirror should really be removed in order to avoid scattering detritus in the tube back onto the mirror. One way is to turn a portable tube assembly upside down. Only do this with the standard cell type, where clips or restraints hold the mirror *securely* in position.[1]

Attempt to blow off any light dusty particles that have accumulated. As with a lens surface, the safest method is to use a jet from an air bulb or ear syringe. An air jet used with the mirror in this position on the mount will insure that most material falls out of the tube. The air compressor can be equally effective, given the preparations used to exclude dust and moisture by filtration, as noted under lenses.

Canned Air: Use extreme care with this product. See the information under cleaning lenses, Chapter 6. Liquid propellant accidentally deposited can badly stain thin-film coatings.

Air compressor: As with lenses, a small electric air compressor of the type sold for airbrushing provides safe and effective pressurized air for optical cleaning. As with all critical cleaning work, you must provide the hose with an in-line moisture and dust trap.

Figured and Flat Mirrors – Advanced Cleaning

A Standard "Wet Method"

Mirrors in open-tube telescopes, as N.E. Howard sagely notes in his classic *Standard Handbook for Telescope Making*, will eventually get dirty enough to require thorough cleaning, no matter how carefully you treat them. The same dusty layers that form on lens surfaces accumulate on mirrors, forming similar patterns along the edges of repeated, dried condensation. As with a lens, this will eventually become dense enough to noticeably scatter light and accelerate any dewing problems. Mirrors don't stand up well to repeated surface cleaning as well as coated lens optics. The (typically) larger surface areas require large amounts of solvent or cleaner, and are difficult to finish-wipe without leaving sleeks in the coating surface. The best treatment is to remove and immerse the entire mirror in a cleaning solution. One could compile a long list of variations, as opticians and amateurs have developed a wide variety of procedures over the years. The method below is tried-and-true, based on field experience.

---

1 Know your mounting cell! Some larger mirrors use slings or unsecured "floating" mounts. Obviously, a tube of this type should *never* be inverted.

The following cleaning techniques work equally well for both figured and flat first-surface reflection coated mirrors. Clean the secondary flat mirror in a Newtonian or the convex secondary in a compound system right along with the primary. The set then gets a fresh start.

Removal of a Newtonian secondary is usually straightforward. Loosen the nuts or fasteners that hold the secondary housing or plate to the spider hub or secondary stalk. Remove the entire secondary unit. A turned lip on the 45-degree angled portion of the housing shell often holds the flat in place, wrapping over the bevel edge. Open the housing by removing any fasteners. Save any packing material behind the mirror for replacement after cleaning.

In some cases, the secondary flat may be adhered to a backing plate, either in the housing or on a mounting stalk. Modern units will generally use silicone sealer; older makers used other cements as adhesive. Such a mirror may be left on the plate for washing. Alternately, a very fine steel wire or thin blade passed under the mirror and carefully worked down will usually part the two. Use the edge of a utility razor blade to remove adhesive or sealer from the mirror back and plate. Do this work on the bench if possible. Re-attach the mirror to the cleaned plate with equivalent dabs of sealer after washing; allow 24 hours curing time, or as directed on the sealer package, before remounting.

Materials List

- Tools required for mirror cell (and secondary mirror) removal
- Pillow or folded blanket to support primary cell
- Sink or tub of adequate size
- Padding fabric to support mirror in bath
- Support block(s) for holding the mirror or mirror/cell assembly tipped for rinsing
- Detergent
- Pure tap water, or filtered water or rainwater
- Distilled water to cover mirror and assembly, plus several gallons (8–12 liters) for free rinsing and decontamination
- Long-fiber surgical cotton, 1 package
- Blotter paper or cotton swabs
- Rubbing alcohol, isopropyl, 91% or stronger, 1 quart (1 liter)
- New household sponge (wet thoroughly, rinse and squeeze damp)

Mirror Removal Procedure A mirror grimy enough to need immersion-cleaning is probably in a dirty cell, too; clean the entire assembly to avoid contamination by dust being carried up from the cell by updrafts after you replace your sparkling mirror.

Compound Systems: There is no general mirror removal method for Cassegrain systems, other compound telescopes, or catadioptrics with accessible primaries; follow applicable manufacturer's directions for removal of the mirror.

Newtonians: With a closed tube, remove the entire mirror-cell assembly from the base of the optical tube. With an open-tube Newtonian (like most Dobs) you will generally have easier access to the mirror, with cell removal optional.[2] With an equatorial Newt, point the nose slightly above the horizon, lock down motions on both axes, and remove the bolts or fittings that hold the cell in place in the tube. Slack the retaining bolts off almost completely in advance so the final removal can be with a few turn of each fastener, and keep a strong hand (or hands – see footnote) against the assembly while working, especially while removing bolts and afterward.

Prepare a folded blanket or other pad, and pull the mirror and cell out directly without wedging. With a tight fit, you may have to use a gravity assist: slew the tube carefully toward the vertical, hold it firmly in place, then lower the assembly carefully onto the pad, which can be placed on a sturdy stool if the mounting is tall enough. Otherwise, put a plastic bag under the pad on the ground and proceed.

Cleaning: The point is to soak the optic as clean as possible before using any mechanical action. Several hours soaking time won't be too long for a very dirty mirror. Some like to leave the mirror on its cell for immersion and clean both at the same time, which works well for smaller mirrors up to 10 inches (25.4 cm) or so. If you suspect there are petroleum lubricants on any cell parts, however, remove the mirror and wash the cell separately.

If cleaning the entire assembly, Loosen or remove any side-supporting grub screws or pressure pads, and wash separately so the cleaning solution can get fully around the edges of the mirror. Slack off the bolts on mirror retaining clips to finger-tightness before immersion, when possible. Leave them in place as a safeguard until the mirror is in the bath, then, remove them to expose the entire surface. Clean the clips with the sponge, rinse in a cup of distilled water, and set them aside to dry. Any felt or rubber padding on them should be bone-dry before you reattach them to the mirror cell. You can use the cell as a holding and drying platform when removing the mirror from the bath. Using the sponge, carefully clean the mirror's edge without touching the coating.

Note: You will be tilting the mirror assembly up *twice* in the sink or tub to drain the surface during the wash process, placing a supporting block underneath the edge to hold it up. If you are washing mirror and cell together, and there is no retaining ring around the edge of the cell, prepare a wedge-block of soft wood in advance to fit between the sink or tub and the lower edge of the mirror disc. Dry-check the fit of the supporting block and safety wedge in the sink or tub before beginning the wash process.

Get the mirror bath ready in a pre-cleaned sink or plastic tub large enough to give several inches of clearance around the mirror and/or cell assembly. If using a garage or utility sink, scrub ALL soap scum, grime, and particles from the sink

---

[2] Get assistance for removal of a mirror assembly any larger than 12.5 inches (32 cm) or so. A large mirror cell is a heavy, awkward load for one person.

and stopper, and thoroughly rinse it down, several times. Fill with tap water or filtered water, depending on your available water quality, to a level that will rise above the mirror or mirror assembly by at least 4 inches. Avoid tap water with very high iron or alkali content, since potentially abrasive particles may be present, or form in suspension when mixed with some types of detergent. Some old hands say to use rainwater; not a bad idea, if you filter it through coffee filters or lab filter paper first. The final rinse consists of two steps, using distilled water. Acquire enough ahead of time to fill your wash tank above the mirror level, plus a gallon or two extra. You won't need distilled water for a cell washed separately.

The safest cleaner is one of the "lauryl" varieties or a pure anionic surfactant detergent such as Orvus WA, available from art specialty suppliers, but any good clean-rinsing detergent without odorants, colorants or skin softening additives works well. The proportion is up to the individual, but a $\frac{1}{4}$ ounce (7 ml) of liquid detergent solution per gallon (4 liters) will cut through most grime. A stronger solution will increase rinsing time. Make sure all the detergent is thoroughly in solution before immersing the mirror. If you are washing only the mirror, place a clean, lint-free white cloth as padding on the bottom of the sink after filling.

Agitate the water above the mirror during the soaking time, without touching the mirror surface. After an hour or so, tip the unit up a little to see if suspended particles move freely off the surface. If not, lightly compact a wad of surgical cotton to the size of an egg, and gently draw it above the mirror surface with a circular motion, barely touching the mirror surface under the water. Tip the unit slightly again and inspect the surface, agitating to move any loose cotton shreds out of the way. Several iterations should suffice. Resist the temptation to rub away at any remaining spots. Leave the mirror alone for an additional hour or two and try again, using a fresh cotton wad. To quote A. J. Thompson: "Surface film resulting from the action of the elements may be present on the mirror, and this can be pretty well removed by rubbing a little more briskly with the cotton *after* all the coarse dirt particles have been washed off."[3]

Use *extreme* care in rubbing the mirror surface in this way, first making certain there are no particles on the mirror surface or trapped in the cotton pad. Oxidation films do occur in polluted or marine-influenced environments, but are less common due to the modern standard practice of overcoating the aluminized reflective surface with a protective layer of silicon monoxide (SiO).

Rinsing: Drain the sink or tip out or siphon off the water from the tub, tip the mirror or mirror assembly to drain it off, then refill with tap or filtered water. Let the unit soak for another few minutes, agitating briskly at intervals to mix remaining detergent and residue, then drain, refill, agitate and drain again. Finally, change the protective cloth under the mirror for a clean one, fill the tub above the mirror's edge with distilled water, and agitate by moving the mirror. Drain the tub again, and tilt the mirror or assembly up 20° or so and place the supporting block under one side to hold it up, wedging the lower end if necessary.

Pour distilled water over the tilted mirror to remove any remaining lint bits or particles. Let it drain clear of any rivulets, then lower it down to horizontal and

---

[3] Thompson, p. 159 – see the Bibliography.

pour isopropyl alcohol over the mirror surface, keeping the entire surface covered by adding a little more alcohol, until the alcohol clears, mixing with any remaining water droplets. Tilt the mirror back up to drain, and absorb any alcohol droplets with the tip of a clean cotton swab stick.

Lift the mirror assembly out of the tub or sink and place it on the support block at a tilt to dry. If washing outdoors, bring the unit inside if possible for final drying.

Strippable Optical Coatings

At the time of this writing, the "strippable optical coating" from Minnesota Mining and Manufacturing, touted for cleaning in amateur literature over the past decade, is no longer available to the average consumer. Non-flexible U.S.P. or medical collodion (typically cellulose nitrate in an ether-methanol solution) is a good substitute if you are used to handling extremely volatile solvent solutions.[4] It is available from art, chemical and medical supply outlets.

Caution: There is no guarantee that any particular mirror has been properly reflection-coated. As with the old "tape test" for solid coatings used by amateurs in shop work,[5] removing the collodion (which adheres much less strongly than tape) can still pull away flecks or patches of coating where it is not securely bonded to the glass surface. It can be argued that a mirror in such a state needs recoating, not cleaning. In fact, even standard cleaning methods will usually remove a coating in such poor condition. Nonetheless, one should go into the collodion process aware of the possibility. The suggestion is not to use it for the first cleaning of a new mirror, after which you should know whether the coating job is sound.

In other words, collodion is not a panacea and its use carries a bit of inherent risk. This said, it has been used to clean some large astronomical lenses and mirrors, and it tends to hold and remove all loose foreign matter from the surface without the abrasion potential of swabs. "Loose" emphasized, because with ether and/or alcohol in a substantially anhydrous mixture, some water-soluble matter or resinous dirt may remain on the surface. In such cases, follow up with standard methods.

---

[4] *Health and Combustion Hazards*: Standard collodion is dissolved in ether (and usually also wood alcohol, to discourage inhalation abuse) – a potent combination that evaporates quickly and requires extremely careful handling. It finds medical application for attachment of instruments and sensors to the skin for diagnostic and testing purposes, so it has passed the basic human exposure regimes in a limited way. However, the solvents can be noxious in closed areas. The concentrated substance is considered poisonous by the US Pharmacopeia (although NIOSH apparently hasn't established a maximum exposure rating).

[5] A small piece of masking tape is lightly laid against the mirror coating near the edge, and pulled away after a few moments. A poorly adhered coating (for whatever reason: poor formulation, insufficient pre-cleaning of the surface) will pull away with the tape. A sad outcome, experienced at one point. Thankfully, the maker replaced the mirror under Warranty, as a defective coating job that was their problem, not mine.

Collodion cleaning procedure: Ideally, the mirror(s) won't be in the instrument during cleaning, but it may not always be possible to remove large mirrors. In this case, be sure to remove or swing away any mirror clips first, to leave the entire mirror surface exposed, and make double sure to level the mirror. One good way to do this with a concave mirror is to place a large drop of water on the center of the mirror surface and tilt the mounting until it rolls to the center-spot (you DO have the center-spot marked for collimation, right?). Alternately, place a bubble level on the mirror cover and level it out.

Before beginning the cleaning, carefully wrap all around the mirror's edge with several layers of strong self-adhesive packaging or masking tape, burnishing it down securely with the bowl of a metal spoon or other smooth surface, leaving an edge of at least several millimeters (1/4 inch) sticking up, forming a "dam" to trap any extra collodion mixture. Mirror-edge accessibility for applying and smoothing the tape edges is a decisive factor for mirrors left in the telescope. Otherwise, don't risk staining adjacent finish coatings just to employ this method. Other methods are sufficient for practical purposes.

Combustion danger! Work only outdoors or in a large, open room with cross-ventilation or an exhaust fan. Allow no open flames, coals, or heated metal elements in the area.

Body protection: Wear work clothing that covers the arms, latex or plastic gloves (collodion is very adhesive and can be a skin irritant) and goggles for eye protection against accidental splashing while working. For large surfaces, wear a well-fitted charcoal-cartridge filter mask of the type effective against paint solvents while working with the collodion. Observe all additional precautions on the container and any package instructions or insert material, including pregnancy warnings.

Experience has shown that a single layer of cheesecloth makes an excellent embedment for the collodion mixture, pulling away from the glass surface smoothly and leaving only minimal residue. Precut a circle of cheesecloth to exactly fit the exposed glass surface.

As described above, level the optical surface and block it securely in place before application. Pour collodion from the larger container into a small glass measure or a plastic dosage cup of the type furnished with alcohol-containing medications and cap the container. A half-ounce (15ml) should be enough for a 6-inch (150 mm) diameter mirror, using a single thickness of cheesecloth. Use proportionally smaller or larger amounts depending upon the surface area of the optic. Be prepared to add enough collodion to completely cover the mirror surface. Apply at once and quickly, within a half-minute or so, since the volatiles evaporate very quickly.

Slowly pour the liquid onto the center of the mirror surface, dispensing just enough for it to spread to the edges of the exposed glass. Immediately center and drop the cheesecloth circle over the wetted surface. Touch any raised areas gently to assure contact, but apply no pressure to the cloth. The collodion will dry thoroughly in a matter of minutes. Check for elasticity or tackiness by touching the bowl of a spoon or other clean object to the cheesecloth surface. The dried matrix will be firm to the touch and have no fresh solvent smell.

When the matrix is thoroughly dry, without touching the mirror surface, grasp an edge of the cheesecloth with gloved fingers or broad-point tweezers and lift it

up slightly. Grasp this edge firmly, and smoothly pull the cheesecloth away from the glass surface in a single motion. Remove any flecks of hardened collodion remaining on the surface by *very* lightly touching them with the sticky surface of a tiny loop of masking tape and pulling them away. Unless the glass was considerably contaminated, it should be sparkling clean. Remove any remaining spots of water-soluble contamination with standard lens-surface cleaning methods.

CHAPTER EIGHT

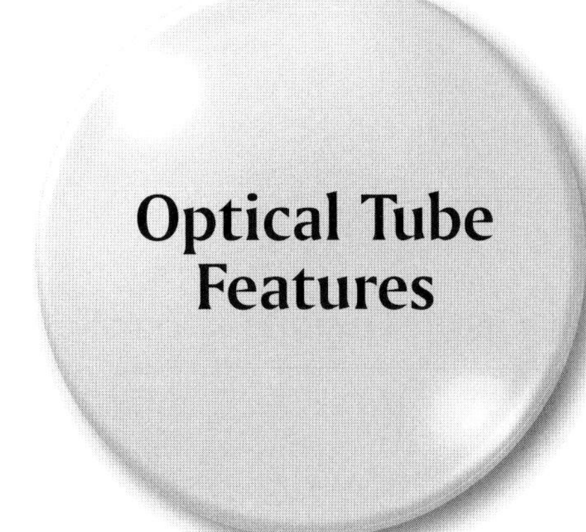

Optical Tube Features

Focusers

Focusers are the one critical moving assembly on most optical tubes. The common type uses a small rack (straight track with teeth) attached to the tube itself, and a pinion (small gear) on a shaft with knobs. The combination draws the tube in and out to focus.

Tension Adjustment

Virtually all focusers have a means of adjusting tension. Adjusting pinion-gear pressure, combined with varying tension on the upper tube-locking screw, will usually take up extra slack or free up a tight focuser. You may be surprised at how much difference small adjustments make, so use trial and error. Rack tension is usually adjustable by turning the four small bolts in the holding plate below the focusing knob shaft. If your focuser is too tight, try backing off the bolts by $\frac{1}{4}$-turn, increasing the amount until you perceive a difference. Tightening, naturally, has the opposite effect, and usually corrects focusing wobble or "run-out" to some extent.

With these assemblies, a finished steel or aluminum cover generally keeps a spring-steel tensioning plate in place, holding the pinion gear against the rack. To see how this assembly works, remove the small screws (usually four) that hold the plate in place. (This is usually part of the routine for installing electric focusing motors as well.) The focusing knob shaft typically has a brass pinion gear pressed on and cemented or brazed in place at its center. Sometimes it is a one-piece machined casting. The spring steel pressure clip may fall or even jump out of the focuser mechanism when you remove the plate. Carry out the operation on a workbench (not out somewhere in tall grass), and keep a hand over the plate, removing it slowly to release tension, remembering which side goes "up." Tip: Don't lose these finely threaded bolts; they are often an unusual metric thread pitch and length, difficult to replace at the local hardware outlet.

Lubrication

If you remove the focuser unit (in a reflector) or the focuser and tailpiece from a small mid-range refractor, you will usually notice that the entire interior of the rack unit is coated with sticky grease. Experienced machinists and ATMs malign this type of lubrication in mounts and focusers, since the high viscosity grease gives the mechanism a tighter "feel" even though the fabrication tolerances may be very loose.

You may opt to clean the entire unit in solvent, including the focuser tube, replacing the lubricant with bicycle grease, as with the procedure for mount bearings. This may or may not be a blessing. Tightening up the action after you replace the lubricant may reveal "bumpy" spots in the gearing that weren't evident before. This is one of many judgment calls in the world of telescope tinkering; the reader is left to make the call.

Interior Finish of the Tailpiece

While the unit is out of a refractor tube, check the surfaces inside the tailpiece casting. Apply flat black finish any bare metal you run into, including the lip (only) of the focusing tube itself – this will help control back-reflected stray light when viewing bright objects.

Some firms have recently upgraded the tailpieces and focusers in their middle and lower range models. Teflon™ shims on cast and machined tracks have generally replaced the oiled fiberboard sleeves or collars that were common until a few years ago. Other improvements include adding a metal plate between the top shim and the tube-locking knob, which allows much firmer adjustment and locking for photography and fixed-focus applications.

With many models, the pressure setting on the focus locking plate and tube shims is also adjustable, by turning small inset-hex or slotted grub screws mounted in tapped holes. The maker doesn't usually cover them, and inspecting the area around the focusing tube aperture will usually disclose the tapped holes. Fine-tune the pressure here by using a jeweler's screwdriver or small hex key. This helps balance adjustments made to the focuser rack-and-pinion. Again, a little adjustment goes a long way.

Friction Focusers

In recent years, Crayford style friction-roller focusers have become common in higher-grade Dobson and third-party supplied focusing units. These seldom require maintenance, other than adjusting the tensioning bolts for the roller assemblies as the parts "wear in." Makers usually supply an instruction sheet showing how to do this In any event, the screws for these adjustments are usually self-evident, usually small nylon or steel inset-hex bolts. Some models have knurled "no-tool" knobs for tension adjustment and lock-down.

Drawtubes

The drawtube in its simplest form is simply a second focusing tube nested inside the main focuser tube. It is fitted to tight tolerances, and allows the extension of focus, usually by 30cm or so, giving a great deal more focusing latitude than the 5 or 6 inches usually provided by the rack-and-pinion or helical focusing arrangements found in the optical tubes of the best modern makers. Once taken for granted even on mid-range instruments, it is now a distinct rarity: so rare that I've seen observers literally jump back in surprise at a user's retraction of the drawtube on a vintage Unitron or Zeiss refractor to accommodate some specialized accessory.

After about 1980, refractor manufacturers in particular assumed everyone would always be using a diagonal with their products, and shortened their focuser tubes by a proportional amount – at the same time doing away with the sliding secondary drawtubes once provided in the tailpieces of all quality instruments. This design change was a sudden, and subjectively reprehensible, step backward in tailpiece design. However, once the drawtube was deleted from standard models and people continued to purchase them, it nearly went the way of the Passenger Pigeon. It costs to add a precision-fitted layer to the tubing assembly in the tailpiece. Exceptions like the 70-mm *Ranger* telescope made by Tele Vue are very unusual.

The aftermarket solution to the resulting lack of focus latitude is fixed-length extension tubes – hard to find for some models, and not quite the same thing, in any case. A variable drawtube accommodates a smooth, analog-variable adjustment for straight through viewing with a variety of prisms, long-focus oculars, and devices such as micrometers, and for narrowband filter units such as T-Scanners and H-alpha filters with specific in-travel requirements. It allows the use of visual telecompressors, offers a built in capability to adjust back-focus for projection in imaging, and the use of remote positions for increasing the power of Barlow lenses for visual applications. Let's bring it back.

Handling and Disassembling Small Objectives

The process of disassembling and cleaning a refractor objective has garnered something of a "secret art" status, along with hand-figuring lenses and parabolization. Few outside of manufacturing deal with the process, and the industry guards its secrets, requiring a return to the factory for many things that hands-on observers would prefer to do themselves. Of course, there are things the home worker can and *cannot* safely do.

Cemented Objectives

If you are looking at a simple cemented objective, such as a finder lens with a dirty or fogged inner surface, there is usually a simple task before you. In any event, it is good practice to take a few older finders apart, clean the lenses and reassemble them before attempting larger optics.

Old cemented glass: Examine for cement defects before trying to make things "better" by cleaning. An achromat with badly deteriorated cement may be cleaner than it looks, since staining or craquelure in the cement may give the appearance of grime. Using cotton-gloved fingers, hold such a lens snugly – top and bottom – during removal. Run a pre-cut band of stiff tape around the periphery, immediately on removal from the cell. Clean only the front and back surfaces, taking care not to let cleaning solution or solvent intrude between the lens elements.

In any case, be careful in handling any older, discolored cemented objective, lest the elements separate, leaving shards of Canada balsam scattered across your worktable. In such a case, you may follow the method (in Chapter 13) for separation and re-cementing.

This should ideally be a planned activity, not an emergency repair! Carefully clean the outside and inside glass surfaces, with the retaining ring and socket, using alcohol/detergent Standard Solution B and rinsing well, as described in Chapter 6.

With a valuable antique glass, you will probably do better by doing little or nothing. Leave the objective in the cell as you found it. Carefully clean the outside and inside glass surfaces, then replace the cell in its tube. Consult a specialist in antique instrument repair for any further work.

Figure 8.1. Using an optical spanner wrench is the safe way to remove the retaining ring from this 50-mm finder objective cell. Photo: M.D. Pepin.

Figure 8.2. A selection of the basic tools needed for disassembling most optical instrument components, including optical spanner wrenches, jeweler's screwdrivers, pliers, and a hex or "Allen" key set for inset-hex head bolts.

A two-element air-spaced assembly will provide a much more challenging project.

Tools: The basic tools required for the work encompass the three main types as noted in the Appendices. Most indispensable for traditionally mounted objectives is the optical spanner wrench. It will be well worth your investment to pick up one of these. *No common tool will adequately substitute for an optical spanner in removing the retaining rings used to hold elements in their cells.* Newer types of cells often use hand-removable retainers. The illustration shows the use of an optical spanner wrench in removing the retaining ring of a large finder objective. Other objectives have knurled rings, which is the case with the 3-inch air-spaced objective shown.

Make a temporary tool for removing smaller rings from a coat hanger or similar stiff wire. Using wire cutters, take out a section of wire and bend it into a "U" shape with pliers Adjust the ends of the "U" until they are separated by the diameter of the ring you want to remove. Trim the ends off evenly and file them down to a blade end (or point) that will fit into the slots or holes in the retaining ring.

Small Air-Spaced Objectives

Figure 8.3. Care must be taken not to scratch the glass in removing the retaining ring from a small eyepiece filter cell using an optical spanner wrench. Photo: M.D. Pepin.

Small objectives may have hand-tightening threaded lens retaining rings. This is nothing new, but it has become more common with recently imported glasses in the lower and middle price ranges. Traditional cells rely on slotted retaining rings that require the use of the trusty optical spanner wrench (or a handmade equivalent) for removal.

The standard disassembly procedure for both is as follows:

1. Remove any external fittings or caps from the objective cell of the optical tube.
2. Remove the objective cell from the optical tube by unthreading or unbolting.
3. Place the cell, front element upwards, on a clean, prepared work surface.
4. Remove any loose material, such as dust or particles, from the inner and outer exposed surfaces using the methods described under light cleaning. Do *not* wet-clean the lens surface.
5. Remove the lens-retaining fitting, usually a ring threaded into the upper or lower cell interior.

Figure 8.4. View through a small air spaced objective before disassembly. Note the three air-spacing shims on the periphery.

Sometimes these are bolts threaded through the retaining ring into the cell assembly. Others, like the example shown, have knurled retainers, removed by hand.

6. Keep the cell in this orientation, so that the lens group will remain in place when you lift the cell.
7. Carefully lift and hold the cell on a prepared soft pedestal so that the lower surface of the lower lens contacts the soft surface. Pause to check the centering over the pedestal, and that the inner cell edges are clear of the pedestal lip. A wooden block or cylindrical container covered with felt works well.
8. Lower or push the cell straight down, uncovering the lens group, until the cell rests on the work surface, leaving the lens elements in their original alignment atop the pedestal

   Caution: If the lens group sticks or wedges in place, do not use pressure, twisting, or sideways motion to try to free it up. Lift the assembly carefully straight off the platform, set it back on the work surface and apply *gentle* opposing pressure. Press on the highest point of the upper lens with a clean, gloved finger or rubber pad, while holding the low point of the lower lens steady with another finger or pad. Alternately, tap the lenses gently, using only finger pressure, while slightly compressing the sides of the cell in several locations. This should free up the lens group. Repeat steps 6 and 7, more carefully.

Figure 8.5. An achromat's cell is lowered to remove it from the lens, leaving the lens group accessible on the padded pedestal. Photo: M.D. Pepin.

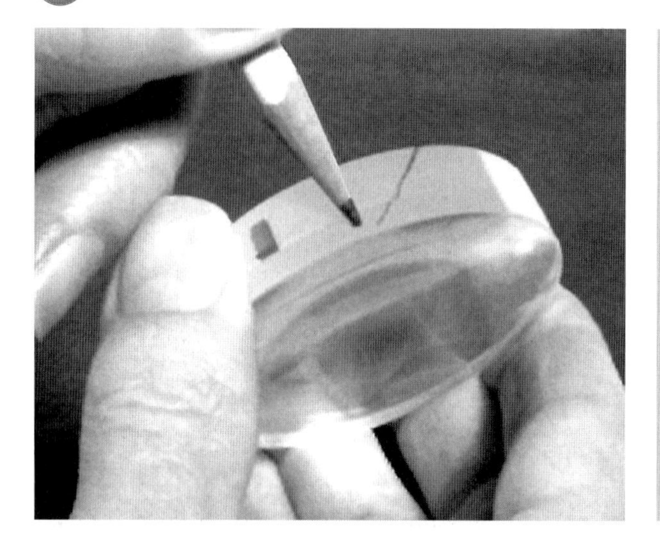

Figure 8.6. Marking the orientation of the lens elements of an air spaced objective with pencil before separation allows accurate reassembly in the original positions after cleaning or maintenance. Photo: M.D. Pepin.

9. Look around the edges of the objective, inspecting for side-taping, an oil seal (after all, you may have an *oil-spaced* objective on your hands) and alignment marks, without disturbing the orientation of the lenses. Resist the temptation to rotate the top lens, should these marks not line up. Let well enough alone for the moment. The elements should be re-aligned *after* cleaning.
10. Inspect the interior facing edges of the lens group for a ring spacer, if no foil shims were apparent when in the cell.
11. If the edges are not taped, tape them together lightly with at three points with small tabs of a tape that will hold securely. Clear Mylar or polyester (not brown) packaging tape has good dimensional stability for the task. Do not use electrical, duct, or cellophane-type office tapes, they are either too gummy or too fragile for the purpose.
12. Draw a set of alignment marks across the edges of any unmarked pair of lens elements, as an alignment guide for reassembly. Make them *easily* distinguishable from any random stains or marks that are already on the lens at a different point on the periphery. Make two or three slanted, close-together, non-parallel strokes with a soft lead pencil, as shown in the illustration.
13. When the marking is done, remove the temporary tape tabs.
14. Test the adhesion of the lenses by carefully pulling *straight* upwards on the upper element. Do not pull hard enough to lift both lenses, and do not use any twisting or sideways motion.
15. If the top lens comes away, lift it completely in one clean motion and set it on edge against a prepared soft resting surface, such as a cloth-covered wood block.

Cleaning of the separated lenses may proceed as in the general cleaning methods described under *Advanced Cleaning* in Chapter 6.

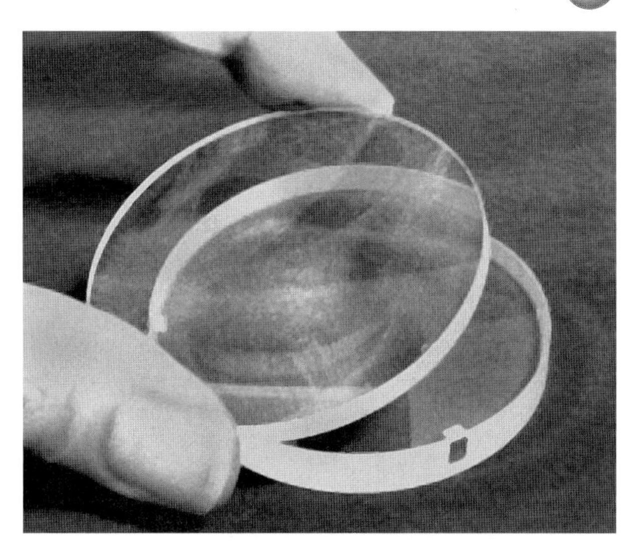

Figure 8.7. After determining that the air-spacing shims are not causing adhesion between the elements, carefully lift the upper element away from the lower to separate the lenses of the achromat.

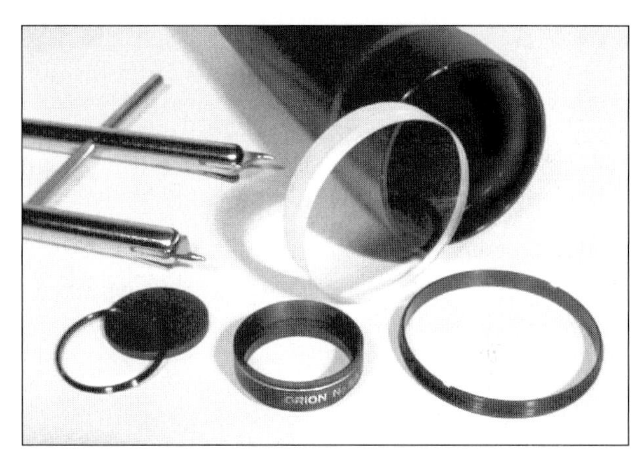

Figure 8.8. Disassembled parts and tools on the bench, showing optical spanner, rings, and tube assembly.

Dew Control and Dewcaps

Strange as it may seem to diehard observers, many people use an outright "dew avoidance" technique. One acquaintance was so frightened by strident cautionary notes in the manual that arrived with his Apo that he simply refused to take any action, for fear of irretrievably damaging the optics. Once the temperature hit the dew point and he noticed condensation on the equipment, he simply packed it in for the night. Others who do this may have started their observing careers in very dry regions, and never developed antidew strategies.

In any case, knowing the mechanism and the safe and available cures, one can rather easily control dew. Remember – you're not trying to keep dew from "falling" on your optics. Dew forms once the surface of your equipment radiates away enough heat that the temperature of its surface falls below the dew point for the humidity level in the air. This is just like droplets forming on a cold drink bottle on a hot day. The concept then, is to slow down this radiative cooling so the optical surface never reaches the dew point.

Removing Dew Using a good dewcap – combined with a miniature hair drier that can be used in an auto lighter socket or auto battery – dewing of the objective and eyepiece can be controlled under conditions of fairly stable temperature and normal humidity (45-65%). Conservative observers overrate the danger of using a hairstyling dryer to remove dew or frost from lens surfaces. However, it does present some hazards to optics, especially to cemented objectives. Another danger, and a habit to avoid, is placing the dryer on the ground between uses. It easily picks up dirt or sand particles on the housing, sucking them through the fan and spitting them out when turned on. Additionally, there is a significant cool-down time after using the dryer, up to ten minutes or more if a large optic such as a Schmidt–Cassegrain corrector plate is heated. Many diehard observers break out the trusty hair dryer anyway, despite warnings, so here is a useful tip.

The cold-air technique: One very damp night, a neighboring observer convincingly demonstrated his little-known method of drying optics: A hair dryer set on "cold" actually removed incipient dew from the objective in a few minutes. I don't claim to understand the physics; possibly the air is heated slightly when it passes the blower motor? No matter, it works on light dew and is certainly worth trying! The cold air technique requires frequent repetition and takes longer to do. However, since little or no heat is applied, there is no cooling down, so the total time spent is about the same as clearing the lens off with high heat.

Resistance-heating: Electrical warming units have been used for years. These use low current passing through nichrome wire of the type used in space heaters, and work very well without the drawbacks of forced air. It would be difficult to argue with the opinion that Kendrick Astro Instruments of Canada developed the first comprehensive system on the block. These and related products are extremely effective, given a good source of electrical power. The method outlined below is a passive strategy for use in moderate conditions, or where all-night freedom from dewing is not a requisite.

Standard Dewcaps: Telescope dewcaps generally come sized for aesthetics and for the convenience of the maker in shipping. For example, a 4-inch (100 mm) objective will ideally have an 8-inch (20 cm) extension beyond the lens, while the standard supplied is usually around 4 to 5 inches.[1] The standard antiglare shields provided with telescopes are also not very thermally effective, especially since they are usually of the same material as the tube itself, and lose heat at the same rate.

Thus, the first line of defense is a protective dewcap, extended to the proper length. Even Newtonians benefit from extension of the tube beyond the focuser. Made in the same way as for any telescope, they also control off-axis stray light

[1] In actuality, some of the smaller, cheaper refractors are made with oversized dewcaps to give them a more impressive appearance; an unintended side effect is better dew prevention!

that can reduce image contrast. A practical cap meant as a dew shield rather than a "glare shield" (which most refractor caps derive from) should be at least twice the length of the telescope aperture. This is the most cost-effective solution for the mobile observer.

Project: When the original dewcap is not effective for at least several hours of dew retardation, the user can easily fabricate a longer one from common materials that have good insulating qualities. Everything from large food tins lined with foam material to carpet-roll cores can be advantageous. The more insulating the material, the more it slows down the cooling of the optics, preventing dew build-up.

In the simplest case, take a full sheet of poster board or thin matboard of a thickness that will not crimp when bent to the curve of the optical tube on which you are using it. Cut it to a width about three times the aperture of the scope. Large sheets of 2-ply artist's matboard are available in 32 × 40 inch size, for instance, and will not crimp when wrapped around a tube of radius as small as 4 inches.

Wrap the entire 40-inch (1 m) length, in layers, but not too snugly, over the largest diameter of the optical tube assembly – i.e. at the existing dewcap or front cell. Apply a strong, clear packaging tape to the outer seam. Slide the resulting layered tube off, and neatly tape down the interior edge. Run a matching band of tape around both ends, crimping it over to seal the ends of the tube, slicing partway into the tape at intervals, so sections overlap and don't bunch up. Then run a band of the same tape around the outside and inside of both ends, smoothing it carefully with a cylindrical object, like a can of soda, to firmly attach the tape to the board.

Spray the entire construction with two coats of a good flat black paint. If you wish, line the interior with self-adhesive black felt, which is very dew-absorbent. Leave a section unlined to slide over the end of the scope. The mounting end will fit snugly, as the tape will take up the slight amount of loose fit you have left to allow siding the tube off.

Voila, instant dew shield for a few dollars and an hour's time!

CHAPTER NINE

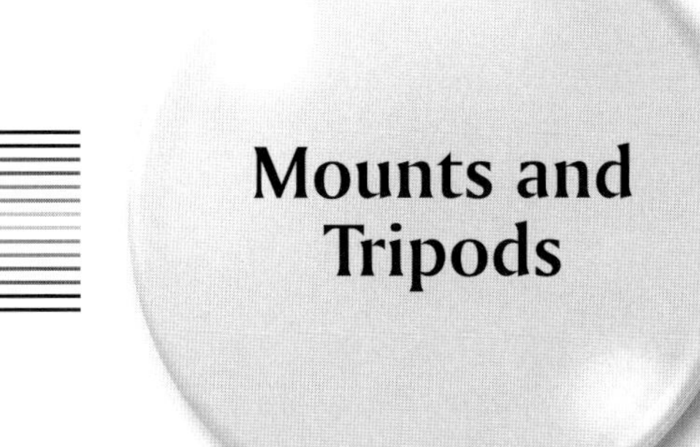

Mounts and Tripods

General Care of Equatorial Mounts

Lubrication

There is little friction generated by the slow motions of most mount and optical tube parts. A thin film of non-acidic oil, metal protectant, or petroleum jelly is generally better than heavy grease on moving parts that may be exposed to handling. Keep any geared mechanisms lubricated as recommended by the manufacturer. Rack-and-pinion focusers and worm gear assemblies exposed to air are points that most often need re-lubrication.

Many modern mounts have the slow motion and drive gearing enclosed in integral housings. These are (more or less permanently) lubricated at the factory. Injection of lubricants or oils of the wrong type may decrease protection by thinning or congealing the existing lubricant. Unless the right lubricant and the proper tools are at hand, refer the work to the maker. The mechanically minded may want to tackle repacking the bearings, as described in the section on Reconditioning Equatorial Mount Bearings, later in this chapter.

Most adjustable fittings on telescope mounts tighten with inset-hex set or "grub" screws. A set of hex keys is an indispensable tool for adjustment work on telescope mounts in general. SAE and Metric keys all have slightly different diameters and there are no exact equivalents. Acquire the proper key set, since undersized keys strip, and slightly oversized ones jam readily (see Appendix A, Hand and Power Tools).

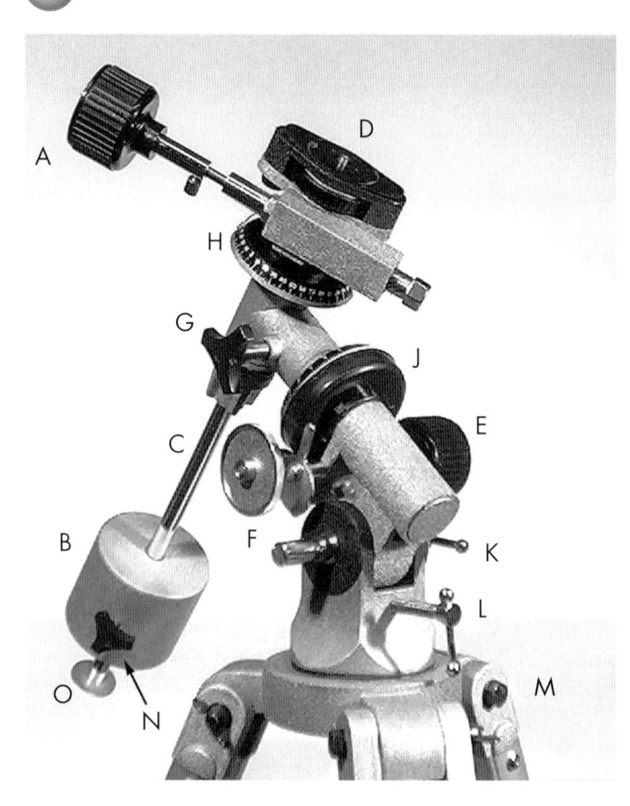

Figure 9.1.
Equatorial Mount–Basic parts and features
A. Declination slow motion knob
B. Sliding counterweight
C. Counterweight shaft
D. Mounting block with attachment bolt
E. Right ascension slow motion knob
F. Sidereal motor mounting stud and gear
G. Declination drag adjustment knob
H. Declination setting circle
J. Right ascension setting circle
K. Latitude tension locking lever
L. Latitude angle setting adjustment bolt
M. Tripod head
N. Counterweight locking knob
O. Counterweight safety stop

Play in Equatorial Shafts

Many equatorial mounts, old and new, have declination and polar shafts that taper slightly where they fit into their bushing or bearing assemblies. One can usually adjust out endplay or wobble by tightening a nut and washer (sometimes a knurled retaining collar) mounted on a threaded portion of the shaft. The collars are usually setscrew-fastened, so use care if resistant to turning—find that screw or screws first! Ideally, do this while the mount is set up, so the unit can be tested and readjusted during the procedure. Polar shaft fittings are generally harder to access than Dec shaft fittings, but the same procedures apply after exposing the shaft bearings or bushings by removing their cover plates or housings. Tip: whether working on the bench or with the mount set up, bear in mind that the result will need to be judged finally with the mount loaded as for normal use.

First, remove any outer housing to expose the nut (or collet). Loosen the nut with the proper spanner or socket, and then tap the assembly down into its bearings or bushings with a non-damaging rubber hammer. A wooden block covered with toweling or rubber sheet will also work well. With the shaft firmly seated, snug the nut down, but not too tight. Back it off and retighten a few times to get the "feel" for the proper tension.

Figure 9.2. A close-up of an equatorial mount head with polar alignment adjustments marked for identification.
A. Polar alignment telescope–focusing eyepiece
B. Latitude adjustment bearing
C. Latitude adjustment micrometer bolt
D. Sidereal drive motor housing
E. Polar azimuth adjustment micrometer knob
F. Latitude indicator plate
G. Right ascension settung circle
H. Polar shaft housing
J. Month and day scale for polar alignment telescope
K. Longitude offset scale
L. Mounting base

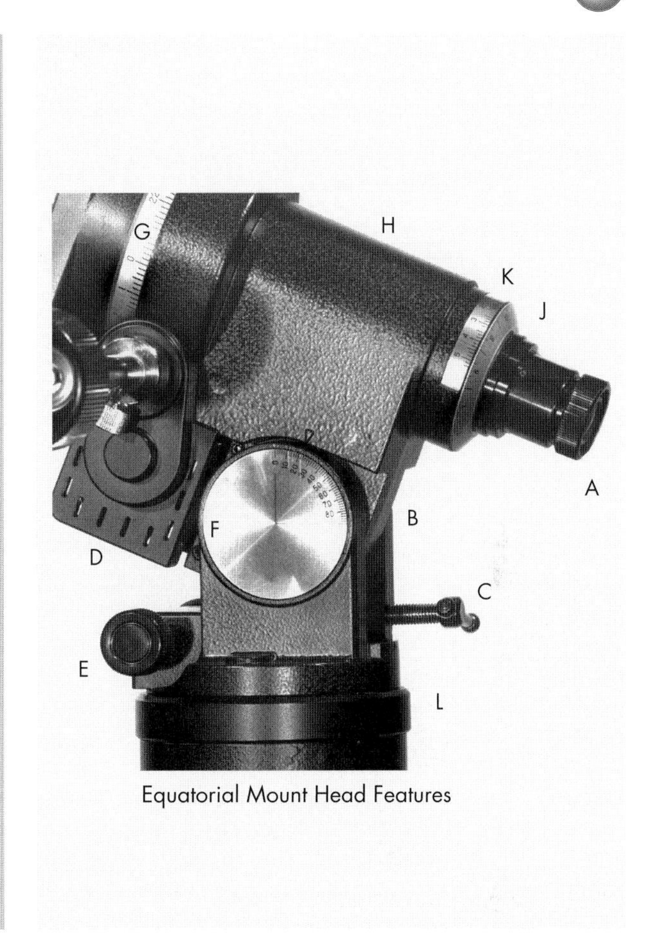

There may be additional metal or nylon washers or other fittings to lubricate. A skim of white lithium grease is a good non-staining lubricant to apply sparingly to such areas. Alternately, if there is a collet on the shaft or lower end of the shaft housing, loosen and move it slightly on the shaft, then retightened to snug up any play.

Repeat the process until the shaft turns smoothly with no binding or wobble. Then attach any counterweights and optics and repeat the testing. If the added weight reveals play in the shaft, run through the steps again to snug up the assembly as well as possible without causing the motion to bind.

Snug up a loose shaft by placing the appropriate thickness of brass shim stock between the shaft and bushing or housing. Shim a single bushing or bearing by fashioning a circular band of a thinner stock so that x/y axial alignment is maintained.

Mounts with untapered polar or declination shafts carried in simple bushings are more difficult to adjust. If one cannot obtain an original replacement bushing, a machine shop or jobber can usually supply or fabricate a close match. Alternately, snug up a worn shaft assembly by installing a sleeve cut from very thin brass shim stock (gauge 0.025–0.075 mm [0.001–0.003 inch]). Unload the shaft by removing any equipment. Then loosen or remove any fittings, pull the shaft slightly upward, and slide the sleeve down between the shaft and bushing with finger pressure. Coat the sleeve first with a small amount of grease, white lithium, or silicone lubricant to ease the installation. The sleeve must be precisely trimmed, and a little undersized. This will assure that it doesn't overlap itself, canting the shaft out of the perpendicular. For a quick fix where there are top and bottom bearing surfaces, place a shim at the same radial location to one side of both bushings, with an end crimped over to hold each one in place.

Loose Tension Adjustment

The type of variable-drag operated by a short lever on a stud or bolt usually has adjustments for slack. Most mounts use the same lever action for both axes. Others, like the small mounting in Figure 9.1, use hand-tightened knobs.

Often an adjusting lever has just worked itself into an inconvenient angle. If a lever is on a smooth or splined shaft, reposition it by removing any fasteners, then pull it off, and replace it at a more convenient angle. The job may require a little careful prying: use a thin hardwood stick, cloth-wrapped screwdriver, or other non-damaging tool. This adjustment will also take up slack from normal wear in the system.

There is sometimes also a tension adjustment nut concealed underneath the lever. The nut may work to hold a slotted bolt head in position. Loosen the nut slightly and retighten by degrees until the action of the lever is smooth and applies sufficient drag. With a slotted-bolt type, tighten the slotted bolt down by degrees with a screwdriver, snugging the nut down to hold it in position. Replace the lever to check, and repeat until the action is smooth and convenient.

Some lever shafts are hollow, with a tension setting bolt visible or protected by a cap. Remove the cap if present. Tighten the bolt until snug, back off slightly and carefully re-tighten. Resistance should vary evenly when turning the lever, not "catch" at a certain point. Put a drop of lubricant or protectant on the bolt head, and then securely replace any caps or other fittings.

Latitude Adjustment

On an equatorial, check the Polar axis latitude adjustment trunnion or knuckle for tightness, and locate any Latitude-height adjustment screw. Loosen the locking bolts or knobs and check to see if the joint moves smoothly, then retighten and check for play. You will need to adjust this setting accurately for Polar alignment to the latitude. Note that many mounts have insufficient tension here, relying chiefly on the weight of the instrument to keep the latitude angle screw in its position against its stop lug.

Even if you can do the work in your home shop, repair and adjustment operations will cost you time. If you still want the mount, a corresponding reduction in

Figure 9.3. Aligning on the celestial pole. Micrometer azimuth adjustment (shown), alternating with latitude adjustment, brings the pole into position in the polar alignment scope's reticle for accurate tracking.

the asking price would certainly be in order. A commercial mount by a good maker may be worth considerable work and renewal of parts. A truly bad mount, however – such as a homemade plumbing pipe design with multiple problems – may not repay the effort.

Drive Motors

Note: This and the following section cover only the commonest problems based on experience with sidereal-drive mounts of the standard type. Most manufacturers of quality equipment have technical staff that will supply owner-tips and maintenance specifications, even for older products. There is also usually practical information in the original manuals for the instruments. When all else fails, try to obtain the manual, or search for someone who owns the same unit; they may have advice on a specific problem.

Drive Gear and Worm gear adjustment: Drive motors on the basic models of both fork-mounted and German equatorially mounted scopes may incorporate a spur gear train (toothed gears only), and a simple AC or DC clock-motor unit for tracking the sky. These mounts are designed for visual observation, and their accuracy over long periods is nominal. Available electronic drive correctors will usually control the rate accurately enough to allow basic astrophotography.

After some time using the instrument you may experience too much slack or backlash in the spur gear drive for your observing comfort. Generally, you can access the drive system by removing a faceplate (with "Cat" scopes), or a drive-housing cover (with equatorials). A spur-gear drive usually has gear positions adjustable for slack, to keep the teeth better meshed with the main drive gear. The general procedure is to loosen the screws on the mounting plate, shift it to take up gear-train slack and carefully re-tighten. Other types have a self-adjusting, spring-loaded meshing arrangement that defeats attempts at user improvement. Most such drives do have a small amount of built-in free-play or "backlash" that cannot be adjusted out without causing binding, excessive wear, or heating up of the drive motor. If you adjust such a unit and find that it drives too slowly, immediately reset the gear position for greater slack and accept some backlash as the price for a working drive system.

Sidereal drives designed for precise guiding and astrophotography generally employ a more sophisticated worm gear drive system, with a quartz-controlled DC/AC system and variable drive rates, generally Solar, Lunar and Sidereal. Backlash and binding are also troublesome in many of these single-axis drives. Some well-designed equatorial mounts with integrated or modular drive gear housings offer no obvious way of adjusting backlash or worm tension without complete disassembly, which may require specialized tools. Refer such work to the factory or to an experienced electrical motor repair shop that works on timing equipment.

Worm gear "Cat" drive bases: After taking the faceplate off, you will find that the drives of non-computerized Cat scopes generally have a spring-loaded worm gear assembly, with the worm on a shaft integral with a motor gearbox mounted next to the main drive gear. These seldom require adjustment, other than checking for well-distributed lubricant, and that the spring is under sufficient tension to keep the worm snugly meshed with the drive gear attached to R.A. shaft.

The one adjustment point is usually a bracket plate that contains an inset grub screw, adjustable against the worm shaft bearing housing to set the worm's tension against the main gear. Beyond very slight adjustment for worm tension, return a malfunctioning drive of this type to the factory for service, since the motor and other components are generally not available for user installation. Although the majority of older SCTs have run for years without an overhaul, a routine servicing of such a drive after every seven years or so is a prudent precaution. While they have unit in the shop, have them replace the crucial drive motor – an easy fix that can avoid an unexpected failure just when you are about to make that 80-minute exposure of the Cave Nebula that's been eluding you for six months due to foggy weather.

Pier-mounted German Equatorial Drives: Simpler worm gear arrangements – such as those found in the drive casings of most pier-mounted equatorials – hold the worm in a "C" bracket housing on a backing plate affixed to the motor-mounting base with small bolts. Oversize holes or slots in the plate allow travel. Adjusting these is a matter of "feel" (and perhaps a thin brass shim-wedge or two judiciously slipped under the plate if the worm has gone off-center).

Generally, the worm gear should move almost imperceptibly before bearing against the R.A. gear teeth, and there should be no "rattle" or backlash apparent at the gearbox when manually budging the R.A. axis. Check the adjustment by slightly pulling eastward and then pushing westward on the counterweight shaft, with the motor off. Repeat this with the motor on, to see how the unit is reacting when you are observing.

In order to control any backlash present, you may be tempted to adjust the side-tension grub screws often provided at the end of the bracket that holds the worm as well. Be aware that the difference between such a drive binding up and its having annoying backlash through "worm wiggle" may be as small as a 1/8 turn of the adjusting screws. Most user compromise, allowing a little backlash, and adjusting the counterweights so that the tube is a little "West-heavy." This keeps the worm engaged so the mount drives smoothly, but lets the drive unit act more as a brake than an engine, thus decreasing wear on the motor and reduction gearbox.

Clutch Adjustment

Fork-mounted "Cat" drive clutches and slow motions: Although generations of amateurs have griped about it, backlash in the R.A. tension and slow motion arrangements on fork-mounted catadioptric scopes seems to go with the territory of nearly all years, makes and models. The clutches are factory-adjustable to some extent, but removing and resetting the angle of the lever on its post and adjusting the R.A. slow motion knob on its shaft are the only user options. As for the R.A. slow-motion knob on older SCTs; putting a thin nylon washer or two underneath, and slipping a short, very thin sleeve cut from brass hobby tubing tightly over the knob shaft to snug up the wiggle in the action can make things feel more positive, but the backlash itself is simply not adjustable.

Significant wear in the system indicates factory return for service. As with drive-base refurbishment, I heartily recommend this, if for no other reason than the difficulty of removing the tapered-bearing shaft nut or other arrangement that holds the fork-arm base to the drive base without losing track of the floating parts underneath. The companies that make these units understand how all that works; they do a fine job of refurbishing their products, although the wait can seem a bit excessive.

Equatorial Mount R.A. Clutches: The variety of clutch types on pier-mounted and tripod-mounted equatorial drives includes everything from stacked, laminated-fiber wheels, to simple nylon or Teflon® washers under tension by a nylon lock-nut where the R.A. axle meets the casting. The main criterion is "if it ain't broke, don't fix it."

However, if the drive is slipping when you add small accessories, then it's time to adjust the clutch. Usually you find a lock nut on a threaded portion of the R.A. shaft, with a second nut or threaded washer underneath that is adjustable to bear on the washer or disc arrangement that maintains clutching pressure. Back off the lock nut and tighten the inner fitting, if present. If there is only one nut, tighten that until you can slew without significant pressure, while the motor maintains a steady drive rate when you change upper end weight by a pound or more. With a two-nut system, leave the inner nut just a bit looser than you want it, since tightening down the outer lock nut will add some pressure.

Tips: With a used unit, a previous owner may have "fixed" the drive in any number of ways, which is why having the original manual for the telescope is an important matter. Factories that are still in business may be able to supply copies, if they no longer have originals on hand. In addition, worn, burred, or bent washers on the R.A. shaft may not seem crucial, but replacing them may fix a jumpy drive.

Lubrication: Non-fibrous automobile bearing grease works well to replace dried out lubricant on the gear and shaft assemblies of heavier scopes in the 10- to 16-inch (26 to 40 cm) mirror-size range, and lighter weight drives for 6- to 8-inch (15 to 20 cm) units shouldn't require more than bicycle-bearing grease, or even white lithium lubricant, on the gear wheels and worms.

Be sure not to put grease or oil anywhere but the places that already have lubrication. You may be tempted to lubricate a sticking clutch, but adjustment is the proper method to cure the problem. Grease or even spray protectant that transfers onto a fiber-wheel or lightweight washer-type clutch may cause problems, requiring disassembly and cleaning with detergent or solvent to cure the resulting slippage.

Setting Circles – Right Ascension

Linked to the clutch mechanism inside the casing, the tension for the R.A. setting circle on a fork-mounted Cat scope is not user-adjustable. A sticking R.A. circle is a factory problem.

With a pier- or tripod-mounted equatorial that has a floating "driven" R.A. setting circle, there is always a "sweet spot" in the adjustment. Some have a threaded adjustment collar or collet, and others combine this with a setscrew that puts tension on a slip bearing around the R.A. shaft. Adjust the circle tightly enough to drive along with the upper portion of the mount, following when you slew smoothly in R.A., but allowing you to reset it without too much pressure. Getting it right may be tricky, but worth it if you want to use the circles efficiently. If you are a star-hopper, don't worry about the finicky adjustment – on less expensive mounts it takes constant attention.

Reconditioning Equatorial Mount Bearings

All petroleum-lubricated bearings eventually need attention. While most manufacturers rate their units as lubricated for life, even in new mounts you will often find a sticky Declination or R.A. axis that needs cleaning and lubrication.

Caution: Resist the idea of squirting household oil or similar substance into the spaces between the moving parts to loosen up the action of a mount. This may work temporarily, but also either dilute or coagulate the grease in the mount, depending on the type used by the maker. Getting to the bearing points, then cleaning and repacking them, is the proper process. The general information below is for renewing the lubrication in a generic cast and machined metal equatorial mount, and does not apply to any specific model.

Tools

Based on the standard toolkit listed in Appendix A:

- Hex (Allen) keys in metric or SAE sizes, fine to medium sizes

- Adjustable wrench ("Crescent" wrench)
- Metric open-end wrenches from 5 mm to 22 mm (or SAE equivalent)
- Cross-point (Phillips) screwdrivers #2 and #3
- Flat screwdriver, medium
- Jeweler's screwdrivers, including flat and Phillips blades
- Large optical spanner wrench (for fittings with slotted rings or holes, generally found at the ends of bearing shafts, or on caps or sleeves)[1]
- *Brushes*: medium-width paintbrush and a small fingernail brush or toothbrush for cleaning parts, gear teeth and bearings.

Materials

- *Solvents*: Mineral Spirits, VM&P Naphtha
- *Rags*: Cotton rags or cleaning rags from auto or hardware supply store
- *Rubbing Compound*: Automotive finish type, coarse and fine
- *Lubricant*: Lithium grease; tube or can for bicycle bearings or general shop work
- *Containers*: Disposable aluminum auto oil-changing pan or oven roasting pans, 2 large jars with lids – quart (liter) size
- *Precautions*: Wear rubber protective gloves and work clothing with a long-sleeve top for any cleanup operations using solvent. Wear safety goggles for eye protection during solvent cleaning. As with any process using petroleum solvents, work in an open space with adequate cross-ventilation or an exhaust fan. Allow no open flame, coals, or heated metal elements such as burners or hot-plates in the area.

General Procedure

Work on a raised surface, preferably a bench taller than table height, with room to spread parts out in order. Use paint thinner (odorless types available) where parts are scrubbed or soaked. A disposable oil-change or roasting pan makes a good solvent tray to hold parts while scrubbing. Use large jars (with lids) for bearings, washers and small fittings free of old grease.

1. Remove the mount head from the tripod head. One can usually do this by removing a hand-bolt under the tripod head, or by using a metric open-end wrench or adjustable wrench to unthread the bolt that holds the mount base on the tripod head.

---

[1] *Note*: As with lens retaining rings, a small make-shift can be fabricated with steel wire bent to a "U" shape, with points or flats filed, ends separated to the distance between the slots or holes. Use it with large pliers on difficult fittings. (See also *Tool Tips* in Appendix A, under Optical Spanners).

2. Methodically disassemble the mount head, first removing all motor housings and attachments, slow motion cables or knobs and any Polar Alignment Telescope fittings.
3. Start looking for the largest bolts in the assembly, and remove them. Most occur in pairs. Work from the top down, being sure to take off any small removable parts or plates that may conceal the fasteners that hold the parts of the mount together. Usually there will be only a few main bolts through the housing, allowing you to split the mount into two parts – the R.A. housing, and the Dec. housing.
4. Remember parts locations by numbering or a short description list.
5. Working in stages, you will eventually find the locking rings, collars, or ring nuts that hold the large parts in place.
6. Most mounts will have plastic or nylon washers between the surfaces to keep metal from rubbing on metal. Save all washers in jars with paint thinner to clean them of residual lubricant.
7. Access to some fasteners may be through inspection holes or by rotating a shaft until recessed holes line up with access slots – look carefully and you will determine where such points are located.

You will soon be down to the bearing surfaces. In mid-range mounts, these are usually not ball or roller bearings, but sleeves or bushings, possibly seating a tapered shaft. Clean each surface carefully, wipe them dry, and set the parts in order on the work surface. Mount styles with non-rotating counterweight shafts may have upper bearings concealed beneath the housings that hold the Declination slow motion arms and/or worm-gear slow motion assemblies. This will be apparent on inspection and removal of the worm assemblies.

Caution: Do *not* use the following polishing techniques in areas with ball or roller bearings installed, as it will wear down the bearing races; just remove and clean such bearings, repack with lubricant, and re-install.

Seating rough parts: Clean all mating bearing or bushing and shaft surfaces thoroughly with rags and solvent. Feel for rough extrusions on mating parts; these points cause friction and binding. Take obvious seams down with wet-or-dry abrasive sheet (150 to 220 grit) wrapped on a section of doweling.[2]

Smooth metal-on-metal surfaces by coating shaft and bearing surfaces with automotive finish rubbing compound. Reassemble without lubrication and rotate the parts by hand, using moderate pressure, until the feeling of "grinding" dissipates. Use two grades of compound: heavy-duty rubbing compound first, polishing compound second. Clean off the residue from the first compound before changing to the second. A few minutes with each compound should smooth things out.

Reassembly: Clean up all parts with rags and solvent. Be sure to remove *all* traces of grit. Dry-buff the parts with clean rags, and use a clean brush and air jet to finish up.

---

[2] A dowel section wrapped in abrasive paper can be gripped in a variable-speed electric drill, and used as a bore-smoothing tool.

Reassemble in reverse order. Coat each moving part and every washer with lithium lubricant. Pack any bearings by pushing grease firmly into the balls, rollers and races. Experimentally tighten the fittings as you go along – don't just reassemble. Press and rotate all mating surfaces; back off and re-tighten collars and nuts on all fitting several times to establish a firm fit. If any part doesn't feel tight enough after assembly, strip the unit back down and tighten any nuts or collars down a little more, then reassemble. The plates that hold the slow-motion worm gears usually have tensioning screws on the ends of the worms. These are a bit tricky, and may require a few iterations to get them tight enough without binding. In the event, a bit of patient working with the various adjustments will result in a smooth-working mount assembly.

Example: The lightweight silver mount pictured in the labeled illustration (Fig 9.1) at the beginning of this section – one example of a type produced over a forty-year period by Asian jobbers for various telescope companies – was rescued from the scrap heap. The motions were sloppy and the Declination axis bound up at several points. Using the above method, it now works as smooth as silk. Even the R.A. circle, inoperative out of the box, tracks accurately with the add-on motor or slow motion knob. The mount easily carries the fast refractor pictured in the illustration of solar filters in the Accessories section, giving good results and high portability for sunspot counts and monitoring.

Aspects of Care of Dobsonians

With Dobs, we are often talking about *big* mirrors by amateur standards, and a lot of weight to manage. The main thread in all Dob work is to keep things as practical, straightforward and reproducible as possible. The standard Dobsonian mount is a lot like a piece of furniture, but with a difference – it lives outside. The philosophy of the Dobsonian user tends to follow practical lines: it could even be boiled down to the prosaic "whatever works, do it!" The wonderful thing is just how many solutions and materials DO work, and very well.

In fact, it seems that every maker, commercial or private, has his or her favorite solution involving unique combinations of materials and novel ways of shaping and assembling them. The specialized literature on the Dobsonian is immense. Much of it exists either in magazine articles or on the Internet. Readers with an interest in the fine points are encourage to browse the many fine magazine articles and websites dedicated to the Dob in all its permutations.[3]

Platform Maintenance and Adjustment

Platform Rotation: The platform base and the altitude trunnion bearings need balanced friction for the mount to function smoothly. The two axes must work

---

[3] *The Dobsonian Telescope* by Kriege and Berry has yet to be surpassed. It covers the topic in general, along with detailed features of specific designs – see the Bibliography.

together in harmony at all altitude angles, something not easy to achieve. One irony is that a big Dob can be particularly hard to "track" when following an object near the zenith, just where the best transparency and seeing is found. This is because the user must impart a slow, even, rotational motion to a resting mass without the natural lever-arm effect that is present when the tube is at lower altitude angles. The designer grapples with avoiding the resulting "hug and twist" routine that can distort the truss tube or upper cage in a lightweight structure.

In general, constant attention to the lower platform bearing materials and tension adjustments is essential. Some of the best mechanical modifications to commercial Dobsonians involve adding (or improving) the provisions for adjustable "lift" in the center of the platform base at the location of the axle bolt or bearing. A strategically placed nut or other adjustable fitting that moves to apply counter-pressure through a plate or washers under the rotating platform works for most users. When properly adjusted, rotating the scope at the zenith becomes much easier.

The problem, of course, is making the adjustments with the optical assembly in place, so that it is possible to compensate for changes in friction due to settling, or temperature and humidity variations, without midnight disassembly. Clamping or welding a lever onto the bearing-tension nut, accessible from the edge of the turntable, is one option. The adjustment bolt or fittings can also be accessed through an aperture or slot provided in the upper turntable.

Controlling weight in the basic design is a good strategy for both transport and handling. Using very rigid base plates, while lightening the rocker box and tube assembly by using thin, strong truss members of materials like graphite composite, or thin tubing aluminum has become standard practice. ATMs often fill the tubing with expandable foam or cover it with pipe-insulation material. This creates resistance to flexure and damps out vibrations. The insulation also makes the metal tubing a lot friendlier to handle on cold nights! Lightweight secondary and focuser cages are commonly fabricated by encasing a rigid framework of wooden ribs with plastic laminate, thin veneered plywood, or other dimensionally stable material. The 18-inch home-designed Dob illustrated in Figure 3.17 is an example of such lightweight, rigid construction.

Altitude Bearings

Makers have taken the opportunity for many design innovations in the important altitude bearings. One classy Dob had perfectly friction-adjusted circle bearings made from PVC toilet-mounting flanges. However ingenious the material used, the goal is creating smooth, controllable motion that won't change with slight variations in top weight. Without friction control, even changing eyepieces – say from a small Plössl to a very large wide-field ocular with a Barlow lens – can cause the nose to drop very quickly, with startling (or damaging) results. Two stock solutions are (a) installing sliding counterweights, accessible from the observing position and (b) varying the friction of the bearing pads in the trunnion base.

Commercial makers originally achieved friction control by altering the size and number of bearing pads, usually made of Teflon™. Many users have modified these systems, including changes to the diameter and surface materials of the side bearings. For instance, using textured Formica™ and other facing materials laminated on enlarged circular bearing faces combines well with the use of trunnion pads or blocks made of material such as Ultra High Molecular Weight polyethylene. The combination has consistent surface "grab," while retaining low wear characteristics. Such strategies help to harmonize the differences in resistance between the two motion axes over a wide range of positions.

Routine Maintenance

It may be stating the obvious, but we do a lot of that in this manual, so here goes: a Dob platform, being right on the ground, is a catchall. Blown dust and vegetation easily collects unseen on the bearing surface of the lower platform. Grit in sandy soils can grind away at surfaces that should be smooth and striation-free for optimal slewing. A slight lip or skirt on the upper rotating platform can help. Nonetheless, carefully clean out the area between the upper and lower platforms and beneath the rocker box after prolonged field use, even if you don't see any accumulation of grit. A few blasts from an air compressor or canned "air," followed by a paintbrush with natural bristles and a damp rag never hurt anything, and will prolong bearing life on both axes.

Damp Intrusion: With the dew a telescope faces in most climates, the components should all be routinely finished to withstand water intrusion. The user should rectify any sealing or coating defects before they cause swelling of wooden components. A good criterion is making sure that the entire structure, exclusive of optics, can withstand overall washing with a wet cleaning solution.

Some styles of commercial mount use "particle board" for the base platform and rocker box. All such components are hygroscopic (moisture absorbent) if not sealed well. Check often for any splits in the facing surfaces. As preventive, it is not a bad idea to run a good stripe of a clear coating (clear nail polish or polyurethane) around the edges and corners of any surface laminate to discourage moisture penetration. Cut edges of particleboard are factory sealed, usually in black. Nonetheless, use a sealing paint in the color of the original and repaint any cut surface not protected by a cap or laminated edge. High quality house paint will serve well, and masking tape will keep drips away from adjacent surfaces.

Lubrication: Unlike an equatorial mount with its metal-on-metal bearing points, few locations on a Dob need petroleum lubrication. However, metal surfaces like the center bolt, any washers or fittings, focuser parts, metallic truss-lock fittings, etc., should be periodically treated with a protective lubricant like WD-40™ and/or a lightweight white lithium grease, removing any excess by quick wiping with a tack rag. Be careful not to change the friction characteristics of any bearing surfaces. Clean any residue away with VM&P Naphtha or rubbing alcohol on a cotton cloth.

Truss Tubes: The truss tube comes in many variations, and the care of them is a straightforward matter of caring for the finish and mechanical soundness of the components. They are generally aluminum or steel tubing. Using anodized tubing is a good strategy for the former. The finish resists weather admirably and only requires occasional cleaning: one of the areas where lacquer thinner gives useful results. Its combination of solvents strips light oxidation from anodized surfaces without harming the finish. Of course, it will attack any paint aggressively, so don't use it on anodized tubing that has been clear-coated with lacquer or enamel.

The steel truss tube may be smaller in diameter, due to its strength, but is subject to the corrosion that attacks all ferrous metals. A little attention with steel wool and naval jelly, followed by rinsing, priming and repainting will be required on a regular basis.

Wooden Tripods

Hardwood tripods are a very attractive and practical feature of portable telescope mountings. Fine makers often prefer the material for its vibration-damping qualities. Although vibration-damping pads can do wonders for steel or aluminum tripods, experience has shown that the same pads used with a wooden tripod achieve superior results.

Mass-produced wooden tripods are also a feature of many refractors and small-to-medium reflectors. Volume makers often seem to have jobbed the work of tripods out to woodworking firms with no direct connection to the activity. Apparently, such makers have no idea of the conditions of use, and simply produce to specification. Taking a generous view, the manufacturers themselves are not cutting corners either, but simply underestimating the frequency and duration of observation. Nowadays, most mass-production firms have gone completely over to metal tripods, and this may soon be a moot point. In any case, there are still many quality wooden tripods under fine instruments at this time. This short section gives a few tips on keeping them sound and fit for use, and gives a few tips on maintaining the not-so-fine ones.

General Care

Protect the original finish on telescope tripods with a solid coating of furniture paste wax, renewed seasonally. Treat "oiled" woods with Danish-style furniture protectant or lemon oil. Like any furniture, the wood surface also needs protection from prolonged sunlight exposure. Reddish stain-finishes in particular are fugitive in sunlight, and will eventually weather to a grayish "barn wood" tone if not protected from the direct sun. This effect is striking when a portion of the leg shaded by the clamping hardware is revealed as the leg is extended to a different position. For this reason it is always best to use an opaque tarp rather than the clear variety for tripod protection.

Touchups: Carry out touchup on scratched or abraded wood surfaces with a simple, museum-recommended remedy. Old English™ furniture polish, a blend of dye and soluble waxes, is an excellent scratch cover and lightweight protectant. It is easily removable from finished wood.

Pigmented wax pencils are handy for deep digs or scratches in tripods and other wood surfaces. The correct shade makes an indetectable repair, sealing

Figure 9.4. Even high-quality wooden tripods with standard wood stain finishes will develop cosmetic fading around hardware if left exposed to UV from sunlight or fluorescent fixtures. Use dark fabric or black plastic for dust covering (when unit is fully dry).

against dirt and moisture. Carefully scrape excess off, going with the wood grain, using the flat of a single-edge utility razor blade. Follow with a very light wipe-down using a pad of cotton cloth dampened with VM&P grade Naphtha or Mineral Spirits. This will level the area, and bond the medium into the surrounding wood grain.

Damp and Dry Conditions: Extended periods of humid weather followed by dry periods should alert the observer to pay extra attention to the tripod. When a tripod has been set up in sunny or dry weather even for a few hours, check the leg-lock tension before attaching the instrument or adding counterweights to the mount. Legs snugged up in the damp nighttime loosen in short order as the wood dries out, sometimes causing a tripod leg to collapse under the weight of the instrument.

Conversely, leg-locks tightened down at the end of a long period of low humidity may constrict the wood as humidity rises, denting it permanently and making the lock less effective at that spot. The wooden part may even splinter under the clamp.

Damp is the worst enemy of equipment in most conditions, yet storage under extremely dry conditions (less than 25% humidity) may also eventually cause permanent shrinkage or deformation of wooden tripods and other wooden parts. A good mean relative humidity (R.H.) level for the long-term storage of wood is 40–50%.

Tips on Protection, Touchup and Refurbishing

Typically, the shaped wood for a mass-produced tripod is dressed and sanded. The worker then sprays or brushes on a staining mixture, rubs it down and allows the piece to dry. Finish coating follows, with varnish or lacquer. Sometimes a simple coat of flat black paint is applied on raw wood. Alternately, varnish-stain is sprayed on, usually a single coat, as with many tripods finished in a light reddish 'mahogany' look. Oil finish penetrates, whereas this minimal varnish-stain doesn't penetrate significantly, and even slight nicks will reveal the soft, bare wood underneath.

Either method is inconsistent with wood surfaces expected to endure extensive outdoor use under conditions of high humidity alternating with rapid drying. Red aniline dye – a chief component of inexpensive wood stains – also fades readily under attack by the UV wavelengths of sunlight or fluorescent lamps.

Refurbishing and Refinishing: Tripods may have faults not readily appreciated on cursory inspection. In most tripods used or examined, the drilling and trimming of the wood to fit in metal caps, etc. followed the finishing process. This is a poor idea, encouraging moisture intrusion and quicker reaction to humidity change. You can check for this and fix it by disassembling the unit, and applying a brush or spray coating of clear synthetic coating to bare wood surfaces you encounter. Run a round brush saturated with coating into any holes that expose bare wood, especially ones that are in contact with metal fittings, since trapped moisture in these spots will corrode the bolts or fasteners.

CHAPTER TEN

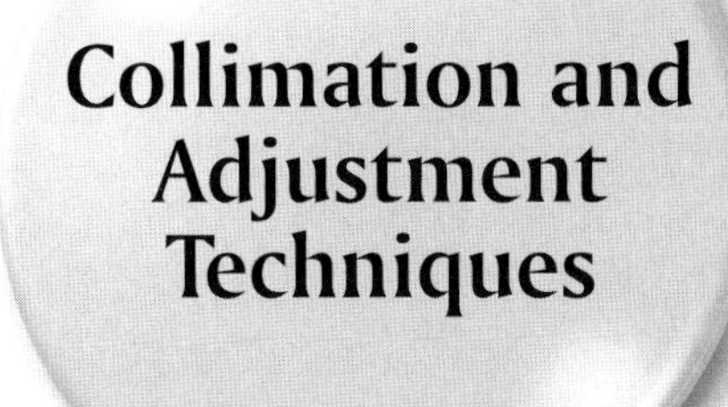

Collimation and Adjustment Techniques

Important Note: The routines given in this section are generic to *types* of telescopes. Manuals or instruction sheets usually cover any special methods and tools for specific models. Always acquire and read instructions for adjusting your particular model if they are available. If advice differs from that given in this manual, please follow manufacturer's recommendations. Some void the Warranty if user adjustments are attempted.

Refracting Telescopes

Routine Collimation Check

There are various seat-of-the pants methods to check for proper collimation. One of the best is the following:

Reflection-autocollimation: This is a fancy phrase for a relatively simple procedure. Obtain a handy-size piece of bright white card stock or thick, smooth art paper. It should be at least as large as your objective aperture, and stiff enough to remain flat when held vertical. Punch or pierce a circular hole of roughly eye-pupil diameter (5-7 mm or so) in the card. A standard office paper punch makes a hole about the right size.

To place the hole at the center of a larger card, cut a piece about an inch (2.5 cm) square out of the center of the large card, punch the hole in it, and tape it neatly back into the larger card. You can also carefully push a circular pencil or tool like a large Phillips screwdriver through the card and clean up the edges a bit. In other words, this doesn't have to be a precision aperture!

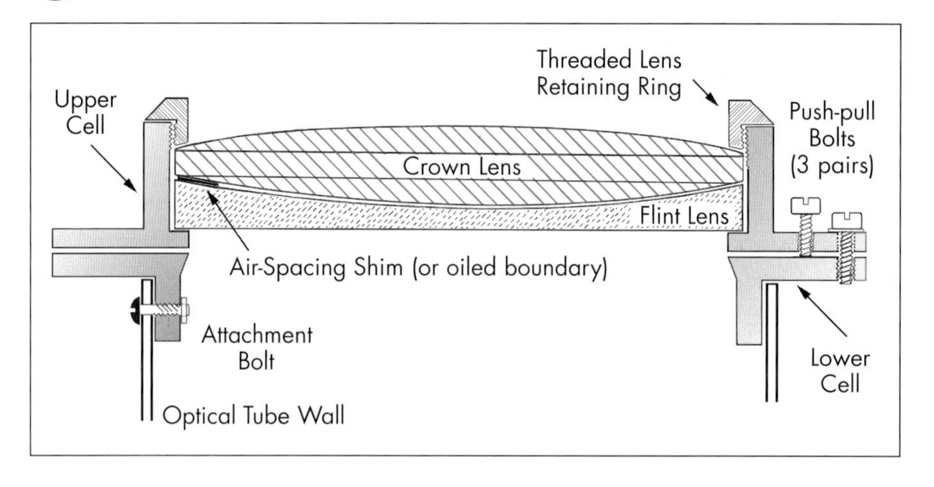

Figure 10.1. Typical adjustable cell for a 2-element objective.

Procedure

1. Set the refractor tube on a level table, or turn it on its mount until it is parallel to the ground at table height.
2. Put the cap on the objective – you want a dark tube interior for this test.
3. Take any eyepiece or diagonal out of the scope's focuser tube. If the instrument has a large-diameter focusing tube, use a 1.25-inch (31.8 mm) or 0.965-inch (24.5 mm) accessory adapter to stop the opening down. This assures you can easily center the card aperture at the opening "by eye."

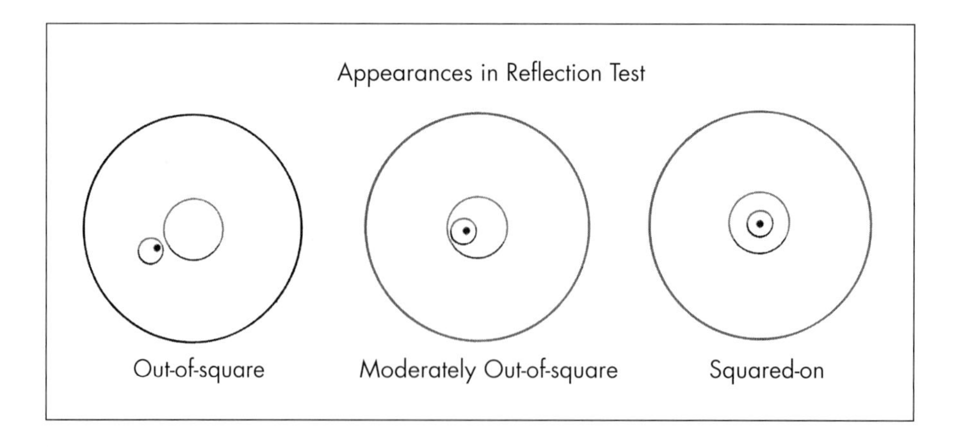

Figure 10.2. Lens cell adjustment.

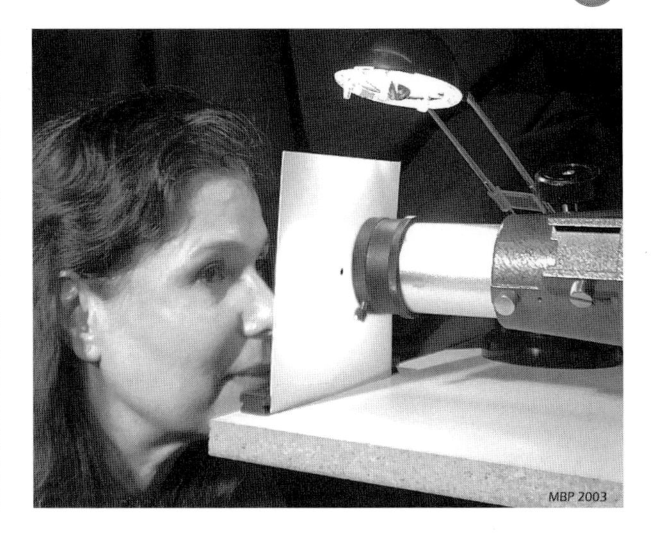

Figure 10.3. The aperture in the card provides a dark central point, also positioning the eye to judge the concentricity of the reflected images returning from the telescope's objective. If the "target" seen is asymmetrical, the objective is not square with the optical axis to the focuser. See Figure 10.2.

4. Hold or position the card 3 to 5 cm (1.5 to 3 inches) away from the open eyepiece holder, and place a bright lamp directly to the side, checking that it lights the scope-facing side of the card, but without casting the shadow of the focuser tube on it.[1]
5. Hold (or position the card on the table surface) so that the card aperture is directly opposite the focuser opening, and look through the aperture into the focuser tube. Exact positioning is not critical, as an off-square (de-collimated) appearance will be apparent without a perfectly centered peephole.
6. You will see several circular, overlapping reflections of the card, one from each lens surface. The dark dot in the center of each image is the peephole and your eye.
7. The illustration shows the appearance of the reflections for in- and out-of-square objectives. If you only see one set of circular reflections, like a small target with a single dark spot at center, thank your lucky stars – the objective is square and all is right with your scope world.
8. If the objective is out of alignment, the appearance of the reflections will be off-center, as in the illustration.
9. Take a pencil and hold it in front of the card, parallel to the ground, with the point intersecting the dark spot-image of the peephole with your eye in the center. Alternatively, you can actually draw a wide, dark line from the hole to the edge of the card with a marking pen. This serves as a reference axis.
10. Note where the dark line you see falls in relation to the direction of the off-center image. Go to the front and loosen one of the "push" bolts on the objective cell about a quarter-turn. Return to the peephole to see whether the

[1] Alternately, do the test in sunlight, turning the mount to illuminate the card. Although you will be pointing at right angles to the Sun, securely cap the scope (and finder, if any). A low Sun is best, since bright overhead illumination can make internal reflections harder to see.

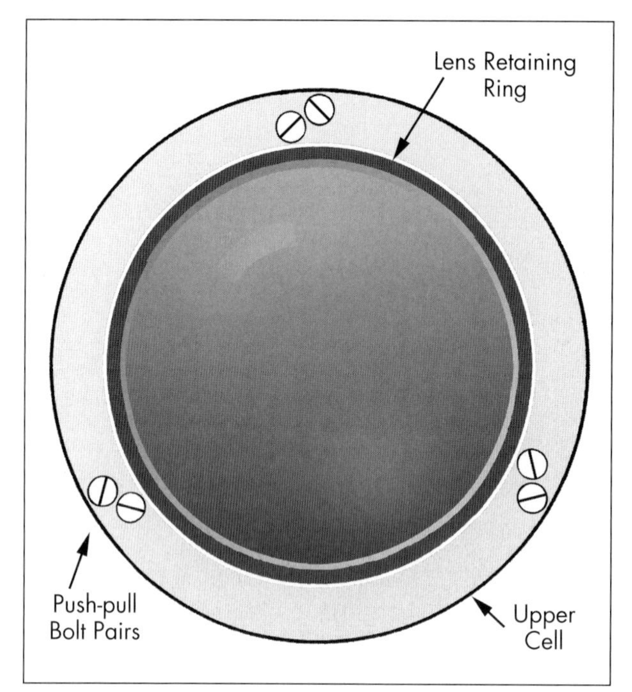

Figure 10.4. Adjustable cell – top view.

images have shifted. If not, go back and loosen the "push" bolt on one of the other three pairs by about the same amount, and tighten the "pull" bolt of the pair slightly.

11. After a few iterations working around the objective clockwise, you should be able to see the overlap changing in relation to your reference line. This will tell you which way to go with the adjustments in order to center up the images and square the objective on.
12. Keep going until the image looks like a single, centered "target." To finish, snug down all six bolts gently and give the edge of the upper cell a few quick taps with your knuckles or a soft plastic object like a film canister. Recheck. If things are snug and the bolts well seated, the image won't change. You are done!

This is a handy reference check to make now and then, especially after the tube has been shipped or through rough transport, so don't toss the card. Stick it in with your telescope maintenance kit (you know; that box with the critical tools and materials you *always* keep with you in the field?).

The "Candle" Test

Another method to check the alignment and optical centering of the lens elements is as follows: Place a small flashlight (or candle, as was formerly done) in front of the open eyepiece holder. An LED observer's flashlight makes a good source for this test.

Look into the focuser opening at the reflections from the lens surfaces. The bright images should cover each other concentrically when viewed straight on, shifting to form a straight line when the eye is moved off-center. If they don't line up, one or more of the following problems may be present: lens elements shifted or tilted in the cell, made of inhomogeneous glass, thicker on one side ("wedge"), or having an offset center of curvature due to improper grinding or edging. If an element is actually out of shape, the objective will never perform well. Reconditioning requires replacement or re-figuring by an expert.

The maker or even cursory factory quality control usually detect badly figured elements, and you will rarely encounter them in a professionally mounted glass. The first thing to suspect is simple loosening or tilting of an element. Removal and resetting of the lenses may correct it.

If the machining of the cell is good, correct tilt by taking up slack in the lens retaining rings, or by readjusting or replacing the foil shims. Be careful to use the same thickness, usually measured in thousands of an inch or centimeter. The spacing ring between air-spaced lenses may be deformed or missing, calling for replacement. Note that few lenses are mounted in contact without shims or a ring. If either are missing, check carefully for adhesive remnants on the periphery of the lens interiors – invariably three patches separated by 120°. In the case of a cell with a misaligned seat, the strategic placement of a thin brass or aluminum shim between the edge of the tilted optic and the seat may rectify the problem. This is properly a job of precision work in a clean shop environment.

If there are no indications of previous means of air spacing, be very careful in handling the glass. You may risk damaging it by trial-run experiments. If it has significant value, take or send it to a working optician to have the curves measured to determine the proper lens orientations. It is even possible someone has cleaned the oil from an oil-spaced objective, for instance, in which case a competent optician can refit and reseal the objective.[2]

Newtonians

Once a scope is really set up *right*, the difference in the image is unmistakable; it's a "whole new instrument." However, many observers groan aloud at the very mention of collimation. The standard visual methods are truly linear and reiterative. If you don't nail the "linear" part, the reiteration becomes interminable.

[2] High-end manufacturers such as Astro-Physics, Baader Planetarium, or Lichtenknecker can perform this work in their shops for objectives of their own, and may contract to do the work for those of other manufacture.

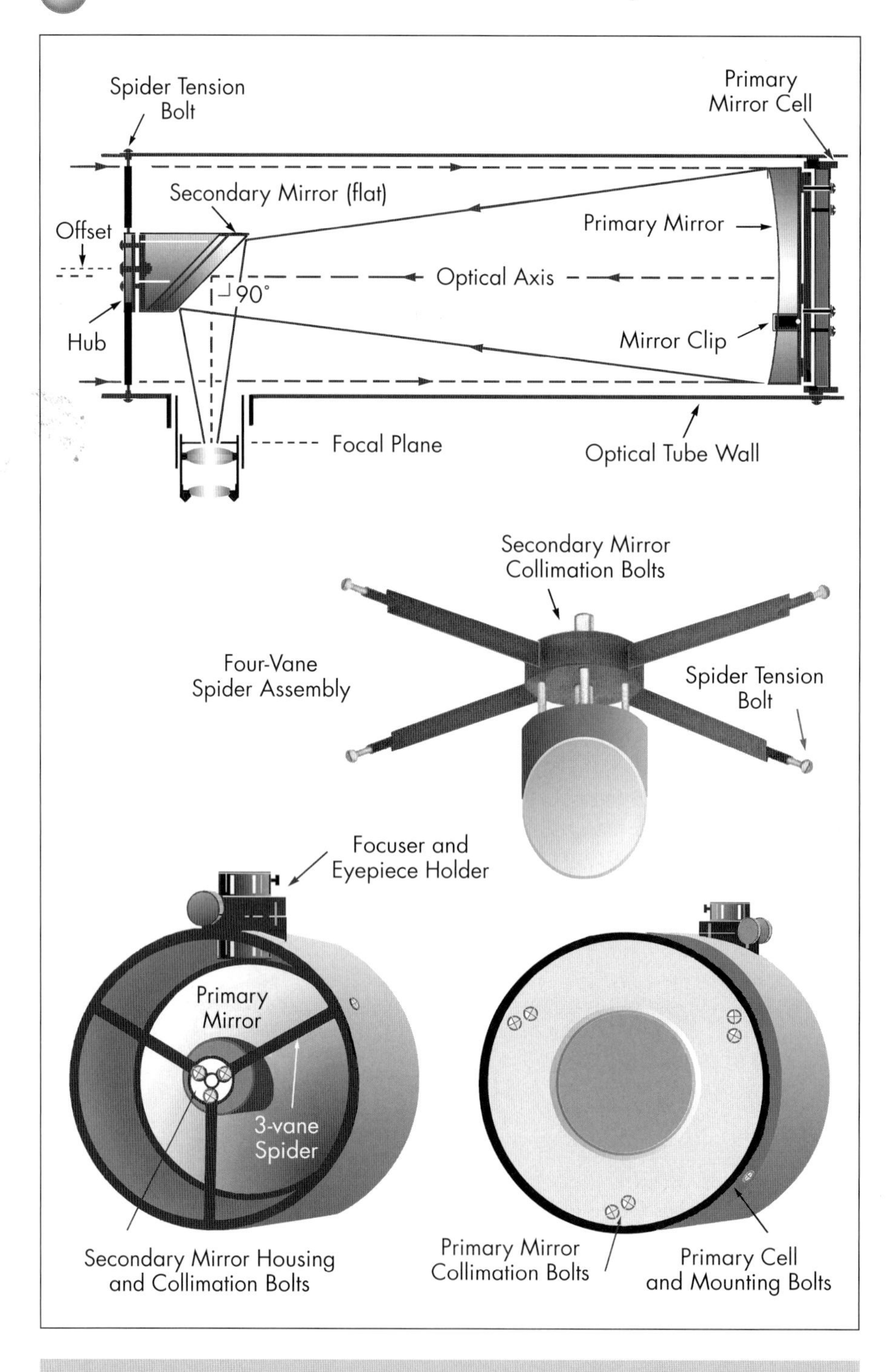

Figure 10.5. Typical closed-tube Newtonian reflector; principal adjustable parts for collimation.

The Laser-Diode Collimator

The arrival of laser collimating units on the scene in the mid-1990s seemed to signal a breakthrough. Early reports were that these pocketsize battery-operated units made collimation "a snap." Given how useful lasers had proven in lining up and measuring things in every field from surveying to aircraft manufacture, this was a reasonable expectation. Instead of spending hours evaluating the confusing appearance of concentric images in a series of eyepieces, we would just point, adjust, pop in an eyepiece, and be off to galaxy-land.

Using one of the first available laser collimators with an Allyn J. Thompson vintage 6-inch f/9 reflector provided an early opportunity to try out the magic.[3] Thompson was a consummate ATM of his time, and used the best of materials. Yet, this solid instrument required considerable tweaking of mechanical structures simply to verify what the laser was showing. It grossly amplified effects from minimal misalignments and minor runout in the helical focuser, not to mention exquisite sensitivity to changes in spider vane tension and diagonal centering. Yet, this was a very forgiving scope. Since the alignment tolerances between focal ratios are (roughly) inversely proportional to their respective cubes, an f/9 Newt gives a lot more room to play with – more than eight times greater diameter of usable field than that of an f/4.5, which is only about 2 mm.

In other words, unless the mechanics are rock-steady, the standard laser collimator is almost too sensitive for use, especially in the long light path of a big, fast Dobsonian. Before the laser appeared, many were lulled into false security – didn't realize that the eyepiece is often constantly moving within the field as the focus knob turns, that the light reflected from primary and secondary mirrors wobbles around due to weight transfer when they are slewed to different positions. However, progress is progress!

Moreover, it wasn't long before people who were used to sight tube and Cheshire eyepiece alignment methods realized that standard adjustment by a naked laser could only accomplish part of a full collimation. Additionally, with the advent of the laser in collimation, the need for a number of adjustments, in many cases even complete overhaul of mirror mountings and tubes, became obvious to the untrained eye. To repeat from p. 36 on performance in Section I: By the strictest definition of a non-aberrated field, the mechanical alignment tolerance of an f/6 Newtonian, considered rather "slow" in these days of large, fast mirrors, is a mere 0.017 inch.... That's a tiny playing field and at f/4.5 it's even smaller. Adjustment on a star image retains its importance under such conditions.

A decade of experience shows that you can look at it in two ways. The tiny beam of coherent light is ruthless, glaringly exposing slight mechanical deficiencies, often adding time to a process that many observers thought they already had a good handle on. Serious use of a laser collimator requires real concentration. On the other hand, hammers don't build houses, people do. Lasers are also excellent for quick evaluations, making a few quick tweaks, and then letting well enough alone.

---

[3] M. Barlow Pepin, "AstroBeam Laser Collimator," *Sky & Telescope,* Feb. 1996, p. 42.

Barlowed laser collimation: Enter an improvement, circa 2000: Given tight mechanics, a less twitchy environment for evaluation is possible with the laser method developed by Nils Olof Carlin of Norway. The concept involves shooting the laser beam through a telenegative lens, which spreads it out, flooding a larger area of the mirror. Instead of working with the tiny, scintillating laser point, you deal with a divergent beam originating at the focal plane. It therefore reflects back a focused shadow image of the mirror's marked central spot on an annular target placed on the face of the lens. This extended image is much less susceptible to tiny mechanical nudges, and readily reveals the basic "rest state" of alignment of the optical axis of the primary mirror with the axis of the eyepiece – which, after all, is the prime criterion for a collimated system.[4] Moreover, adjustment is possible while the target is projected, without the constant danger of a coherent beam striking the eye.

With a typical Newtonian, mirror alignment isn't the starting point for collimation in any case. One thing that experienced observers seem to agree on is that one should start from the outside and work in when collimating any telescope that allows adjustment of mechanical tube components. Check, straighten and tighten all relevant parts before attempting direct visual or laser collimation. This is the only practical way to set up for aligning the mirrors themselves.

Mechanical Adjustments – Spider Hub and Focuser

Menard and D'Auria describe some useful tools and methods (see the Bibliography). The procedure given here is indebted to their advice and to that of other experienced observers over the years.

Materials list

- Tools for removing the secondary mirror housing, adjusting the spider-vane length and tension for your model of telescope (screwdrivers, hex keys, open-end wrenches, etc., as required)
- **Note**: See manufacturer instructions for any special procedures or tools needed
- Hardwood dowel stock, 1.25-inch (32 mm) diameter, a foot or so (30 cm) in length
- Hardwood dowel stock, $\frac{1}{2}$-inch (~12.5 mm) diameter, a little longer than the scope aperture
- Electric drill with bits up to dowel diameter, table vise with padded jaws
- Brass or plastic (Mylar™ is ideal) shim material slips if necessary, up to about 1–2 mm thickness
- A piece of clean, straight wooden lath or flat molding stock, about the length of the scope aperture
- Soft pencil with eraser, for marking.

---

4 "Collimation with e Barlowed Laser," Nils Olof Carlin, *Sky & Telescope*, January 2003, p. 121.

Procedure

The object is to adjust the focuser perpendicular with the optical tube/optical axis. Otherwise it moves *across* the focal plane when focusing, displacing the optical axis of the eyepiece relative to the "sweet spot" at the center-of-field at focus. As a reference point to check this, we first adjust an equally important setting: the secondary's radial symmetry with the primary's ideal optical axis within the tube or truss. With a truss Dobsonian this means fully assembling the instrument, so you can work between the upper cage and mirror box. Mentally substitute "upper cage" for "tube" throughout these procedures, if relevant. Have the optical tube fully assembled, and *remove* the secondary mirror and housing.

Note: The first procedure, centering the secondary or "spider" hub, is relevant to any compound telescope (e.g. Classical Cassegrain, Ritchey–Chrétien) that has a spider-mounted secondary mirror.

Center the Secondary Hub

1. With a plain piece of lath or flat wood molding stock, measure from the tube edge to the center hole or bolt of the hub from each spider vane attachment point; trace a pencil line on the back of the stick where it crosses the front edge of the tube. This will show any de-centering. Alternately, measure the inside lengths of each spider vane to an accuracy of a millimeter or so.
2. *Adjust the hub position* if necessary by alternately loosening and tightening the spider vane fasteners. Loosen one, then tighten the opposite (or adjacent vane, in a three-vane system), and check with the measuring stick. Continue working around the perimeter, maintaining vane tension, until the hub or bolt centers in the tube. At this point, the measurement marks on the stick will overlap (erase superseded marks as you go, so they don't build up and confuse the issue).
3. *Vane tension check*: Make a sonic double-check by carefully tapping each vane with a screwdriver handle and putting your ear close to it. It should emit a low tone, nearly the same for each vane, and you should hear a note, not the dull "thunk" of a slack vane.
4. Be careful not to distort the tube by over-tightening the vanes. Make it nicely circular by marking the outside of the stick and sweeping it radially around the center point when you are finished. Take note of any lengthening of the tube radius between the vane attachment points. Serially loosen the vanes to correct this, and then recheck each vane's tension.

Secondary offset: I'm not trying to steam up anyone's reading glasses, but in my opinion you probably don't need to "offset" a typical Newtonian secondary's position; just make sure the vane lengths are equal, centering the hub (and definitely *center* a Cassegrain's hub).

The secondary mirror in a manufactured Newt is nearly always sufficiently oversized to insure that offset is not necessary to capture the entire light cone from the primary. Conversely, the diffractive asymmetry resulting from an offset secondary is so slight that it has no discernible ill effects, so offsetting can do no harm. A centered secondary, however, allows you to use "perfect" symmetry as a visual check for collimation, whereas an offset secondary gives a slightly de-centered appearance when properly collimated.[5]

Check the Focuser Alignment Make a tool for this by fabricating a wooden plug that fits snugly in the smallest eyepiece holder you normally use, and about the length of a standard eyepiece, circa 2–3 inches (25–50 mm). Of course, 1.25-inch (32 mm) hardwood dowel is a natural for this. Drill a hole straight through the center of it, to accept a dowel rod of nominal $\frac{1}{2}$-inch (12.5 mm) diameter.

- Start by trimming the ends of the dowel section square with a miter box or table saw, and mark the centers of the circular ends.
- Rough-sand or shave down the finished piece if too snug a fit in the focuser; alternately wrap it in clear packaging tape (well burnished down) to take up any minor slack.
- Hold the plug in a table vise, jaws padded with cardboard slips, or use a tubing adapter if you have one.
- Drill holes, one from each end, to meet at the center. Start with a 1/8-inch (3 mm) bit, and increase the size by steps to match your dowel. A drill press will do a neater job.
- Check that the hole is parallel to the plug sides by putting the dowel in place and rolling the plug on a flat table – a slightly off-center hole that is parallel to the plug sides will still work, but be sure you align the de-centering *directly* fore or aft in the focuser when you do your centering checks.

You can have this piece made by a millwork shop, too, but be sure to give them the dowel for test fitting if you do. Make sure the focuser is tight by checking and snugging down the attachment bolts.

1. *Place the plug firmly into the focuser* and rack it until the focusing tube is around the height for observation with a mid-range eyepiece.
2. *Slide the dowel* downward through the hole, stand in front of the tube and sight on its position as it crosses the spider hub. If it misses the center at all, the focuser needs radial adjustment. Now, check the longitudinal centering by measuring the distance from the center hub to the dowel, compare it to the measurement at the top where the dowel enters the tube (the stick-and-pencil method will work fine here, too).
3. Rack the focuser in and out, noting whether the rod tilts back and forth. Extreme change means you need to adjust the focuser rack tension.[6] In any case, do the adjustments here with the focuser tube in the middle of the range

[5] If you know that your secondary is precisely sized to capture the cone with no room for error, you may wish to offset the secondary position. Formulas and online computer programs are readily available to determine this "downward" adjustment of the spider hub. The amount is generally less than a millimeter or two, even in a large reflector.

[6] See Chapter 8 for the procedure. If this is ineffective, you may eventually want to replace the focuser unit.

you would use for observing. If this all comes out dead-center, thank your lucky stars! Otherwise, follow the next steps in the procedure.

4. *Adjust the focuser position if necessary*: First try tightening one or another of the focuser mounting bolts. You may be able to center up the rod by doing this, avoiding the need to shim the focuser body.
5. *If this is ineffective*: Loosen the bolts that hold the focuser to the main tube until it rocks slightly. Gather your shimming material. Sighting on the rod, test-fit shims under the focuser body plate until you determine at what position and how thick a shim you need to center up the rod on the spider hub. Test-tighten the focuser, and readjust the shims as needed until the radial sighting and longitudinal measurements match up.

You now have a squared-on focuser and a centered secondary hub. Remove the test rod and plug, replace the secondary mirror on its housing in the hub,

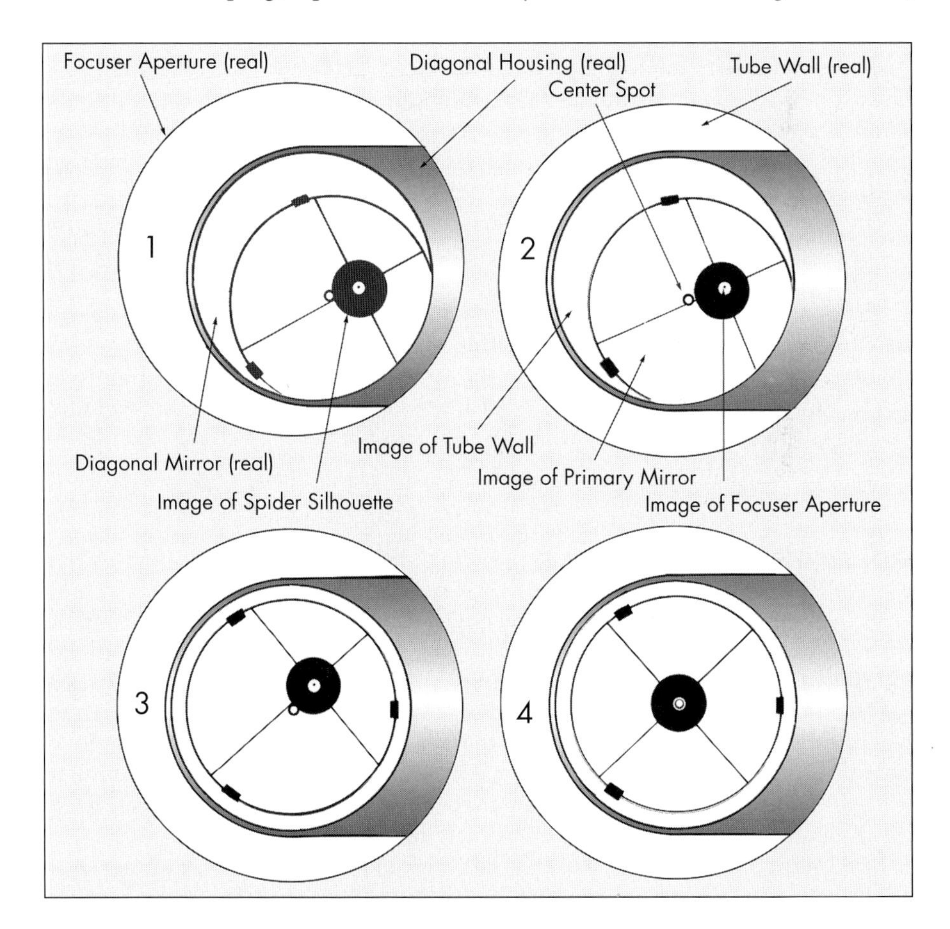

Figure 10.6. Stages of collimation in a Newtonian reflector. 1. diagonal decentered in aperture, all axes noncoincident; 2. diagonal longitudinally and radially centered; 3. image of primary mirror centered in diagonal; 4. collimated–primary axis coincident with diagonal axis.

and tighten the attachment bolt just snug enough to allow the housing to be turned.

Visual Collimation

Special tool: Take a plastic 35-mm film canister and cut the bottom off flush with a sharp knife or hobby saw. Then mark the center of the cap and pierce through at that point with a needle to mark it. Heat up a large nail, such as an 8-Penny size, holding it with pliers in a flame or on a hot range burner. When it barely begins to glow, let it cool a few seconds, then push the nail through the cap exactly in the center. This is your collimation peephole tool – indispensable.

Note on Collimating Eyepieces and Lasers The cross-hair eyepiece, Cheshire eyepiece and the Autocollimator eyepiece are all excellent tools. There are several sources including Tectron of Sarasota, Florida who first re-popularized the tools in the 1980s. The makers can provide special instructions for their use.

The laser collimator is also an excellent device, and its use is a specialized procedure with its own literature and various suppliers. All are good and recommended for use by experienced workers.

The procedure here, however, is purely visual and uses only the simple peephole to center the eye. The notion behind this bare-bones method is to get the user truly familiar with the working of the optics, able to achieve a close approximation to collimation by eye alone. Someone who has mastered the skill will then be able to further hone ability with the use of any of the specialized and worthwhile tools noted above.

Adjustment Procedures *Looking into the tube*: The view into a Newtonian focuser can be disorienting. Try to ignore the reflections, just look at the real secondary housing and the real secondary mirror. This is hard to do through a peephole, so leave it out for the first steps. Look down into the focuser tube aperture at the secondary housing, and turn it until the bright surface of the real secondary mirror appears circular. Make this easier by placing a white sheet of paper on the bottom of the tube below the secondary, so your eye has some revealing contrast and a reference point for the real tube wall. Move your head slightly to get a sense of the center of the focuser tube, while identifying the real parts, versus their reflected images.

Bear in mind that example No. 1 in the *Stages of Collimation* illustration shows a Newtonian that is further out of adjustment than you would generally find a scope just arrived from the factory to be. If your scope looks as symmetrical as example No. 4 to begin with, then you don't need to do the *Visual Collimation* routine at all. You can go right to *Collimation by Star Image* to determine if the mirrors need fine adjustment.

However, if you have already taken the time to do the mechanical checks and adjustments outlined above, it won't take much time to go over a "dry run" of the *Visual Collimation* steps at this point. It will familiarize you with the adjustment parts, where they are all located, what tools you need, and so forth. You will know where to start if the mirrors are ever knocked out of alignment in transport or decollimate in the normal course of use.

Adjusting the Secondary Mirror

1. After the mechanical adjustments you have done, it should be easy to center the secondary top-and-bottom within the real focuser tube aperture, as in example No. 2 in the illustration.
2. The secondary housing, however, may not be centered from front to back (right is "front" in the illustration). If not, adjust the bolt or other fittings that move the secondary longitudinally in the tube, until the real mirror appears circular and centered in the real focuser aperture as in example No. 2.
3. Insert the film canister with the peephole in the focuser, and look through the hole at the secondary again. Your eye is now centered in the tube, where the optical axis of an eyepiece would be. Make any slight adjustments to the secondary necessary to center it up top-to-bottom and right-to-left. Remember to ignore the reflections: look at the *real* secondary mirror and housing while you finish this step.
4. Now consider the reflected images. The image of the primary mirror may look somewhat off-center like it does in example No. 2., with the reflected image of part of the tube wall showing as a crescent to one side. Ignore the silhouette of the spider. The dark outer circle of the primary mirror, its central spot and the mirror clips, are the points of reference for this adjustment.
5. If the image of the primary now seems off-center top-to-bottom, rotate it secondary slightly until the image is centered top-and-bottom.
6. From this point on, you are trying to adjust the image of the primary so the outside ring, mirror clips and center spot look like they do in example No. 3. Again, ignore the spider silhouette – you will center it up later.
7. The primary's image may still be off-center left-to-right. The only way to adjust this is to alter the *tilt* of the secondary by adjusting the three small tilt-adjusting bolts (or whatever type of adjusters your spider has on its hub – it is usually three small inset-hex bolts, or Phillips-head bolts on the front of the secondary hub).
8. Since you already have the image centered top-to-bottom, this is the time to tighten down the nut or fitting on the secondary, to snug it up and keep it from rotating while you adjust the tilt. Don't make it "monkey-tight," just snug it down enough so it doesn't rotate easily.
9. Examine the secondary hub again, if you haven't already identified the tilt adjustment bolts. One of the adjustment bolts on the secondary hub will be at either the top or bottom of the hub. It is usually at the *upper* (focuser) side of the hub. Adjust this bolt first.

10. Loosen it very slightly, while looking through the peephole. If the image of the primary moves farther off-center left-to-right than it already was, retighten this bolt to a snug point. If the image of the primary is *still* off-center to left or right, you will need to loosen both of the other two (usually *lower*) bolts on the hub, each by the same amount. This will change the tilt of the secondary further in the same direction it was moving when you snugged up the first bolt.
11. Loosen first one, then the other bolt by a very small amount; even 1/8-turn usually tilts the secondary significantly. If the image of the primary moves in the right direction, you are just about done with this part of the adjustment.
12. Tighten the first bolt you adjusted again – there should be a little slack in it now that you have loosened the other two. If the secondary moves *too* far (remember, you want the primary circle and clips to center up as in example No. 3), stop, and loosen it by 1/8-turn or so. Go back to the other bolts and tighten them both by a slight amount.
13. Reiterate this process, tweaking the first bolt, then the other pair, until the image of the primary with its mirror clips centers up as in example No. 3.
14. Finally, tighten down the central secondary housing nuts or bolt, and tap the secondary housing lightly to see if it goes off-center. If it does, the fasteners are not tight enough. Carefully recheck tightness, taking the time to get it right.

You have now squared the secondary mirror with the mechanical and optical axis of the telescope.

Adjusting the Primary Mirror

Center Spotting the Primary

If your primary mirror doesn't already have a center spot applied by the manufacturer (many newer scopes do), remove the mirror cell and mirror according to the procedure in the manual, or follow the directions here under *Advanced Cleaning.*

Find the center of the mirror: draw a circle of the exact measured diameter of your mirror on wrapping paper, or any sheet larger than your mirror. You can use a nail and a string with a pencil held upright, or a large drawing or carpenter's compass. Cut the circle out precisely as a template. Lay this template carefully over the mirror, center it up and put the point of an indelible marker through the center hole made by the compass point or nail, lightly touching the mirror surface. This will create a tiny center-spot.

Now, take a single self-adhesive hole- reinforcement of the type used for reinforcing the pages of three-ring binder sheets, and carefully apply it to the marked mirror center, with your tiny mark exactly centered in the hole. This creates a visual reference in the "shadow" of the secondary mirror that has absolutely no effect on the visual performance of the telescope, but is a wonderful centering mark for collimation. Reinstall the mirror.

Alignment Procedure

It helps to have someone to work with you to adjust the primary. It can be done solo; it just takes more time. With a scope larger than about 6-inch (150 mm) aperture, you won't be able to just look back and forth to make the adjustments, but will have to step back and forth, from adjusting one of the mirror cell push-pull bolt pairs, to the eyepiece to check the effect, and back again to tweak the adjustment.

1. Assess the look of the scope interior with the film can peephole out of the focuser. The only thing obviously out of center will be the spider silhouette, with the small, reflected image of the focuser tube at its center, although this will darken or perhaps disappear as you move your eye or head closer to the eyepiece holder. The bright center will darken or disappear when you put the peephole back in the focuser.
2. Concentrate on the position of the spider hub silhouette and the center-spot annulus. The goal is to get these centered in the field of view.
3. At the mirror cell, loosen the adjustment nearest the focuser tube (the "top" adjustment), and check the image of the spider silhouette and center spot. If loosening moved them further off-center, you are going the wrong way.
4. Keep your eye on the mirror's center spot. The goal is to center it up on the reflected image of the primary, centered right over the tiny bright hole that appears if you back up from the peephole a little and examine the reflected image closely.
5. Reverse the process as needed. To do this you will need to loosen the opposing fittings, and tighten the ones you are working with.
6. Keep going, slowly working around the cell and checking the effect of each adjustment. The appearance in example No. 4 is your goal, and is not that difficult to achieve. The spider vanes will appear of equal length, the tube wall will disappear from view, or appear as a dark, even annulus around the image of the primary, and the center spot of the mirror will be reflected in the center of the diagonal shadow.

This is as close to true collimation as you can get with the simple visual peephole method. The scope should be perfectly usable as it is, but you may want to perfect the mirror alignment. The star testing described in the second section below will take you from here, to a truly collimated state.

General Daytime Tests

These are helpful for assessing the state of a telescope in general – *not* rocket science, but interesting to try, and revealing of any real problems.

You can make an approximate daytime test for general resolution using a small, detailed graphic as a test target. It should include straight lines; printed characters and bright primary colors – a color page from a magazine, a colorful banknote, or a playing card will serve well.

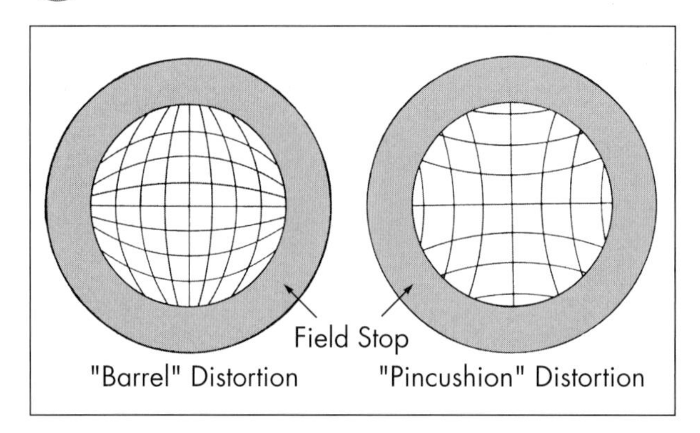

Figure 10.7. Distortion of a grid at focus.

It is useful to have two identical test targets. Compare one at hand with the view of the distant object through the optics. View the target at a distance of 25 meters or more, at successively higher powers. Breakdown of the image under high magnification – changes in color, color fringing, distortion of straight lines, or the lack of a flat field – can all be detected in this way.

The distortions known as *pincushion* and *barrel* are easily detectable by viewing a sheet of graph paper at a distance of 50 feet or more at moderate magnification (see illustration). These common effects, most often from a wide-field eyepiece, are hard to detect while viewing the sky at rest. When moving the telescope across a star field, however, the pattern of stars appears to expand or contract as it comes into the field of view. Narrow-field oculars such as good orthoscopics generally correct for this factor, which can be crucial when directly estimating or measuring exact distances.[7]

Professional optical test-pattern charts are also available through scientific suppliers. These can give a qualitative assessment of such factors as astigmatism and resolution. The main benefit is the ability to roughly measure resolution under the conditions of observation by noting the smallest visible separations between calibrated lines. Extreme astigmatism is detectable by variations in the density of line patterns seen at different radial angles (if in the observer's eye – always a consideration as well – the density will appear to change when the head is rotated while looking in the eyepiece). Detailed evaluation of impressions from such a test, however, is the province of the expert.

[7] Bear in mind that barrel or pincushion distortions – straight lines curving slightly at the edges of the field – are among the least compromising of image-plane defects for astronomical use. Due to the nature of light, no design can be perfect, and this distortion is sometimes abided in favor of better corrections for the more significant defects of spherical and chromatic aberration, coma, astigmatism, and higher order Seidel aberrations.

Collimation by Star Image

Star Testing in General

Star testing is the most unforgiving examination given to optics, and the few instruments that achieve a near-perfect score are world class. Recognition of the exact nature and extent of problems is a skill gained only through experience with telescopes of varying quality. Nonetheless, truly compromising defects will show their presence at first sight, and this is the chief value of star testing in its simplest form. If the defects are still glaring after one has eliminated factors such as poor collimation, the observer's eyesight and aberrations introduced by accessories such as a star diagonal or defective ocular, then the system needs attention.

For instance, a star image that stretches into a line when slightly out of focus reveals a problem. If the line changes orientation by 90° on either side of focus, the system has astigmatism. Geometric distortions, such as oval or triangular star images, are a bright red flag to the observer that one or more optical elements in the instrument are out-of-figure or subject to warping under weight or pressure. In short, any fixed deviation from radial symmetry of a focused star image or defocused Fresnel "target" image of a star held at the center of the field of view indicates a compromised factor somewhere in the optical system – usually a fault in manufacture or mounting.

Typical Airy disc and Fresnel ring appearances in the focused and extra-focal star images of a good, unobstructed optical system were illustrated in Chapter 3, Figure 3.13 and are repeated in the comparison diagram in this section. The evaluation of deviations from these ideal diffraction patterns forms the diagnostic procedure of star testing. The character of the star image at focus; the pattern and consistency, the width and brightness of the Fresnel ring array as it expands on either side focus all play their part. This takes the star test out of the realm of "snap judgment," because there are so many factors that influence the appearance of both the disc at focus and the rings when outside of focus. In this book, we leave applications outside of collimation to the working optician or optical expert, to whom one should refer an instrument with glaring anomalies.[8]

Collimation Using a Star Image

Caution: Be sure to read the Important Note at the beginning of this section regarding manufacturers' recommendations before attempting to collimate your telescope by star image. Specifically, consult the available literature about the adjustments for *your* particular model, which may not follow the generic method given here.

---

[8] For a complete description of the process for a number of optical systems, one can't do better than Suiter's comprehensive work, *Star Testing Astronomical Telescopes* (see the Bibliography).

Star collimation requires decent seeing, sufficient time and the proper tools. It is a process of fine-tuning, generally reiterating the mechanical processes done with the unaided eye, but on a level of finer discrimination. The image in a system that is *far* out of adjustment is bound to be asymmetrical, and thus difficult to correct on a star. The method assumes that the user has already collimated the telescope within rough parameters. With refractors, this means that the objective has been squared-on within the loose tolerances of the reflection-autocollimation test using the standard mechanical/visual methods described in the section on *Refracting Telescopes* at the beginning of this chapter. In a Newtonian, the observer has adjusted the mechanical relationships: diagonal centering, longitudinal and radial adjustment, and alignment of the primary mirror.

With most compound telescopes – classical Cassegrains for instance – the primary and secondary mirror tilts are the usual accessible points for user adjustment. Commercial Maksutovs, in general, are factory pre-adjusted and the user can't collimate them. Makers of other catadioptric systems generally recommend the star method as a first resort. With SCTs, for instance, secondary tilt is generally the only adjustable parameter and is amenable to adjustment on a star image.

Atmosphere: The atmosphere is a major factor in collimation, since the wavering of the star image under atmospheric or local air disturbances easily confuses adjustment. Nights of very good "seeing" are uncommon in most climates, and one usually has to settle for a decent approximation, say an Antoniadi level II or III (two common Scales of Seeing are in Appendix B). In any case, use good judgment in location, avoiding local seeing "traps" such as viewing over cooling road surfaces or near chimneys and heated roofs, and let the optics settle down to the thermal mean of the season before proceeding.[9]

Choice of star: Plan to keep the same star centered in the field of view for some time. Ideally, the test star should be around 40° from the horizon or more, to reduce air-mass-induced seeing effects. The region between the Zenith and the Celestial Pole is a good choice in middle latitudes, since the star's R.A. motion will be slower. When using an altazimuth or Dobsonian mounting, point to the western horizon and slew up to a star that has already crossed the Zenith. This will align your azimuth axis for nearly straight-line tracking as you follow the star smoothly down the sky. If time goes on and it gets too low, just hop up to another similar luminary.

Visual magnitude (Vmag) 5.0 to 6.0 stars work well optically in most amateur class telescopes, avoiding optical scattering or irradiation in the eye from a brighter star. Moreover, finding and re-acquiring the star will be easier with the unaided eye (or through the finder where skies are less dark). Stars of white through yellow color (Spectral Class O, B, A, F, G) will work best, generally the "whiter" the better. Avoid stars that are visibly "red" (K or M class), since the images are subject to physiological or optical "bloating."

Mounting and motion: A solid and well-adjusted motorized tracking mount is a great help. With practice, one can move an altazimuth with slow-motion gearing to follow the star with precision. A Dobsonian or simple altazimuth can

[9] Winter star testing is rough in northern latitudes, and the generally turbulent skies recommend themselves better to deep sky work.

be "nudged" along to do the job. With large telescopes, it helps to have an assistant. Some ATMs have added adjustment rods accessible from the eyepiece position. In any case, all collimation demands reiteration on several levels, certainly not a novel situation to the keen observer!

Appearance of the Star Image: The observer is concerned with meeting only one set of criteria. The final goal is obtaining a good, *symmetrical* star image at focus, with the central Airy disc surrounded by faint, circular diffraction rings. Obstructed telescopes will present the central obstruction as a *dark* central disc when viewed out-of-focus – with obstructed systems, this is the locus around which to adjust out-of-focus symmetry.

Note the star images in the accompanying diagram. The top pair shows the star image in a *grossly* de-collimated, unobstructed system. Far subtler deviations from symmetry will have deleterious effects on the image at the focal plane. Most states of de-collimation will more nearly approach the images at the bottom of the illustration. More than one adjustment session under a steady atmosphere are usually required to achieve this final condition.

Tip: Where the collimation adjustment points are fitted with inset hex bolts, setting three snug-fitting keys of the same size, one at each adjustment point, will assist in the work by reducing fumbling for slots in the dark. Some observers replace the slotted or Phillips-head bolts in their collimation adjustments with inset-hex head bolts to facilitate this. The hex key tips can be coated with a hard adhesive, such as a layer of epoxy, to achieve a snug fit that will keep the keys in place for collimation, yet allow easy removal.

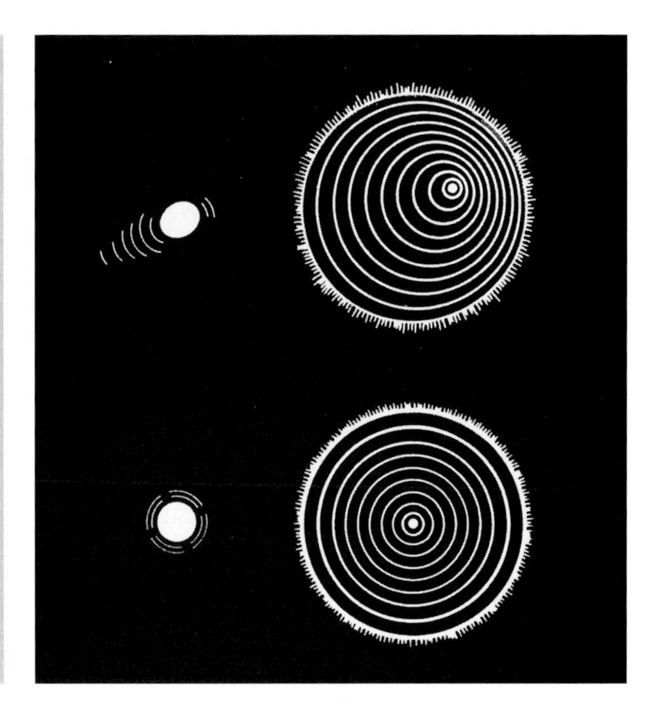

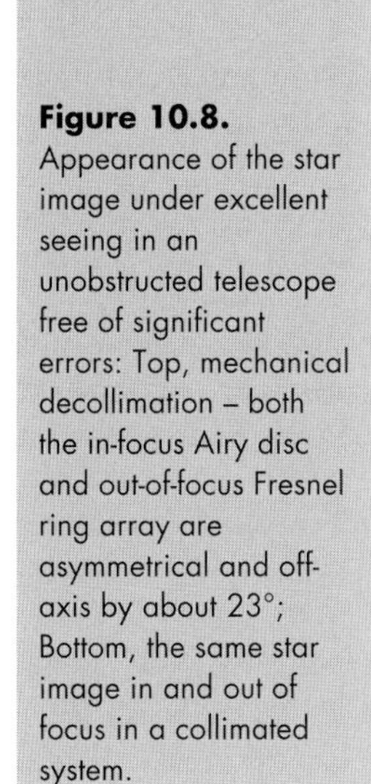

Figure 10.8. Appearance of the star image under excellent seeing in an unobstructed telescope free of significant errors: Top, mechanical decollimation – both the in-focus Airy disc and out-of-focus Fresnel ring array are asymmetrical and off-axis by about 23°; Bottom, the same star image in and out of focus in a collimated system.

A General Procedure for Primary Mirrors, Objectives and Secondaries

Refractors There are three points of adjustment on the objective cell, as illustrated in the section on *Refracting Telescopes* at the beginning of this chapter. Have your screwdriver or Allen key(s) handy. The procedure is one of reiteration of very fine adjustments.

Newtonian Reflectors As with visual collimation, you will be adjusting the three push-pull collimation points on the mirror cell. The procedure, as for the refractor, is one of reiteration of very fine adjustments. Always recheck the secondary mirror adjustment before proceeding.

Schmidt–Cassegrain and Classical Cassegrain Secondaries

These have similar adjustments for the secondary housing cell: on the spider hub of the Cassegrain, on the front of the secondary housing at the center of the corrector plate for the SCT. The difference between this and the refractor and Newtonian is that here are (generally) no push-pull bolt pairs. The procedure, again, is one of reiteration of very fine adjustments. The whole process consists of serially tightening and/or loosening the three small, equidistant adjustment bolts that control the tilt of the mirror.

The manufacturers describe the locations and access for the three inset-hex, Phillips-head, or other bolts used for adjustment. Most also give cursory instructions in their manuals. The instructions here are generally applicable.

For obstructed systems, follow the same reiterative procedure as for a refractor. Unlike the refractor, the Fresnel array will have a dark central spot, giving the whole system the appearance of an annulus. The goal of achieving a centered, symmetrical pattern is, essentially, the same.

Important: In all systems, the tightening of one bolt must be relieved by the reciprocal loosening of one of the others in the pair or trio, or the entire fitting will be stressed, with possible damage resulting.

With SCTs and Cassegrains, do *not* adjust the central bolt that holds the mirror plate unless manufacturer directions specify doing so for some particular part of the operation.

Checking for Decollimation: Bring the chosen star to the center of the field and keep it there. View the image both in- and out-of-focus, scrutinizing the image at both positions for asymmetry. If none is detected at 50× or so, boot the power up a little and check again, taking the process right up to the usable magnification limit of 50× or more the aperture in inches (2× in millimeters). If you detect definite asymmetry at any stage, begin adjustment.

A general principal is to find the adjustment point that moves the star pattern *towards* symmetry when it is tightened. This is your principal adjustment point. Tighten it while loosening the other two points, keeping the array centered laterally on the line of motion caused by adjusting the principal point. Go a little beyond symmetry, and then gradually tighten the other two points by degrees to bring the Fresnel pattern back to symmetry. Keep the secondary shadow or central star image at the exact center of the field of view. The higher the magnification, the more accurate the adjustment will be. Observers often use shorter oculars than they would normally observe with, when adjusting on a star. This will be effective for almost all systems with only slight decollimation.

Many procedures state that the first bolt pair to try on a Newtonian is the pair that is "on top," or nearest the focuser. This is for convenience only, and has no practical relevance, since it may not be the principal adjustment point.

Specifics of adjustment:

1. Loosen one of the adjustment fittings from a sixteenth to an eighth of a turn and note whether the Fresnel ring pattern has shifted, and in what direction. If it doesn't change, loosen one of the other adjustments by about the same amount until you find the one that brings the Fresnel pattern towards symmetry.
2. With bolt-pairs, tighten the "pull" bolt of the adjacent pair slightly after loosening a "push" bolt.
3. Working around the objective or mirror, make tiny, serial adjustments, checking on the star image after each iteration.
4. Keep the star in view and as closely centered as possible with the slow motion controls; the Fresnel array will begin shifting at some point. This will tell you which way to go with the adjustments in order to go beyond center from the principal adjustment point, and come back with the other two points to square the optic on.
5. Work slowly and methodically. Be careful not to throw the system out by over-adjustment of one fitting.
6. Keep going until the image looks like a single, centered "target." To finish, snug down all bolts gently and give the edge of the optical cell a few quick taps with your knuckles or a soft plastic object like a film canister. Recheck. If things are snug and the bolts well seated, the image won't change. You are done!

Miscellaneous Adjustments

A Note on Tilted Component Telescopes

The tilted component telescope was briefly described in Unobstructed Reflecting Telescopes in Chapter 3. One can't ignore the fascination of designs that many are convinced rival the best refractors in image quality. At this point, however, the author must emit a cry of "uncle," and refer readers to the specialists for further practical information. Most professional optical manuals have nothing to say on the subject, and personal experience is limited to a few sightings and a single session using one.

Other than plans for making the Tri-Schiefspiegler and the Yolo-style reflector – which can also be had custom-made at a price – there are virtually no U.S. mass-manufacturers of Brachytes, Schiefspieglers, Medial Refractors, Herschelians, or other unobstructed off-axis systems. According to European and U.K. astronomers, they are somewhat of a rarity outside amateur opticians' circles there as well. This is a shame, as interesting as these types are to the theoretically minded observer and amateur optician.

It is possible, however, to include a short note from limited personal experience. As mentioned above, users claim that the images in good TCTs rival those of the best refractors. There is no doubt that they can. Several years ago, the chance came for a long look through a TCT under good atmospheric conditions. This well-adjusted 5-inch (125 mm) Tri-Schiefspiegler yielded a tantalizing similarity in image quality to a 5-inch Apo refractor on the same observing field, although it was difficult to make an exact judgment in terms of contrast. Given the difference in focal length between the f/10 Apo and the f/34 Schief, it also took some juggling of eyepieces to get the same magnification worked out! However, the 5-inch optical tube alone weighed about 50 lbs (23 kg). The overall bulk was easily twice that of the refractor, and in an odd shape for transport. This was a specialized instrument made by a dedicated and fastidious observer.

TCT Collimation: One often hears the comment that these are "forgiving" systems. Certainly, long-focus spherical mirrors are more forgiving of slight deviation than short-focus paraboloidals. However, a look at the published specifications of one such design revealed that claims for ease of collimation might be a bit exaggerated. The specifications listed tolerances for mounting the mirrors were in terms of a few hundredths of a degree per surface. This is not trivial precision. Admitting that each element has a more relaxed adjustment tolerance than the typical Newtonian primary, the challenge is the sheer number of surfaces, up to four in newer designs. This begs the question of how much time one would spend compensating for accumulated errors in collimation.

For the Schiefspiegler in particular, one would like to be able to refer the English reader to the specialized works of designer Anton Kutter of Biberach, Germany for practical advice. Unfortunately, although the centenary of his birth was celebrated this year, the bulk of his writings remain un-translated.[10] The handful of abstracts turned up in English includes the 1960s articles by noted American ATM Oscar Knab and master optician Richard A. Buchroeder. These don't reveal much about the adjustment and handling of these instruments in the field. It remains for an enterprising optician to widely re-publish the practical use of these excellent designs, as incentive for their manufacture and appreciation.

---

[10] Kutter, A., *Der Schiefspiegler* (Biberach, 1953), and several articles in the publication *Sterne und Weltraum*.

Diagnosing Diagonal Alignment

Most users employ a star diagonal for all their visual observations with refractors, "cat," and compound scopes. Improperly aligned or figured diagonals are often the source of problems imputed to the scopes. Some major manufacturers have gone to prism diagonals as stock accessories due to recurrent quality control problems with small flat mirrors manufactured in bulk. For this reason, always remove the diagonal before attempting any optical testing: test the diagonal while you are at it.

To check for problems, remove the diagonal and center the scope on a target about 50 meters distant, focusing it in the crosshairs of a low-power cross-hair eyepiece.[11] Tighten the mount's locks. Now, replace the diagonal without shifting the tube, and refocus.

Note that the image will probably have moved, partly due to recentering in the accessory holder, partly due to focuser or mirror shift, and partly due to inaccuracies in the diagonal. Such factors are endemic to interchangeable accessory systems. The newer holder types that use expansion rings do a better job of centering accessories than the standard setscrew holders.

In any case, re-center the image, and rotate the diagonal, after barely loosening it. This takes a bit of neck-craning, but you will be able to tell easily whether the diagonal itself is out of alignment, since the image will wander symmetrically around the field of view as you rotate it. If the change is noticeable, you should consider swapping or returning the diagonal, since most are not adjustable. You might try careful disassembly of an inexpensive unit, to check that the prism or mirror is not simply tilted a bit in the housing – otherwise acquire a replacement.

Aligning Optical Tubes in the Tube Cradle or Mount Block

All mounting systems are prone to optical tube maladjustment through mechanical flexure, irregularities in manufacture, imprecise mounting, or post-manufacture shifting of the fittings. A common fault of equatorial mountings is a slight misalignment of the optical tube, where its position in the cradle or on the mounting block is out-of-square with one or both axes. Although it won't interfere with star hopping, this problem is so common that nearly every equatorial mounting I have used needed some adjustment for this factor before even simple setting circles could be used with repeatability. A more serious difficulty is that misalignment in the cradle or block introduces a systematic error that impacts tracking precision, computer or setting-circle pointing, and long-exposure imaging. The situation affects computer pointing because the software points the mounting, not the tube itself. The digital "learning" routines that compensate for period drive-gear error can't correct for it, since the computer routines assume

---

[11] If you don't have one, you can make one at home – see Chapter 13, *Making Finder and Eyepiece Cross-Hair Reticles.*

an orthogonal setup. Dovetail-track mountings can alleviate much of the mechanical problem, but are not immune.

The situation is easy to detect with computerized mounts, since the user programs the error in when centering initialization stars in an off-square optical tube. These pointing errors will persist, despite re-initialization or resetting on new alignment objects over the course of a session.

The situation may require constant attention to visual guiding (or more frequent auto-guider corrections) because the optical axis is cumulatively non-orthogonal with the Polar alignment or the guide stars used for initialization, the extent of "wander" depending on the hour angle and altitude of the initialization objects and the area being subsequently tracked or imaged.

Adjustment Routine: We take a tip from surveyors here. Place a low-to-moderate power cross-thread eyepiece in the main tube to center a reference point on the optical axis. Any ocular with a marked center will do. A simple crosshair eyepiece like the commonly available 20-mm Kellner suffices. Photographic guiding eyepieces with micrometer reticles are also ideal.

A simple non-illuminated cross-hair eyepiece is handy for so many observing and adjustment applications that every observer should consider keeping one around for general work. If you don't have one, you can easily create your own, as described under *Equipment Projects and Tips* (Chapter 13).

General setup: Whether observing or adjusting, the tripod should always be firmly seated, on antivibration pads if available, and allowed to settle into the ground surface before adjustment or alignment is undertaken. With light-to-moderate weight mounts, shake the entire assembly lightly and tamp it down a bit with gentle blows of the hand. Also, re-check the tripod or pier legs for tightness before proceeding.

The procedure is as follows:

1. With the telescope set up as you usually do, critically align the finder with the main tube, centering a distant, fixed mark in the main tube and adjusting the

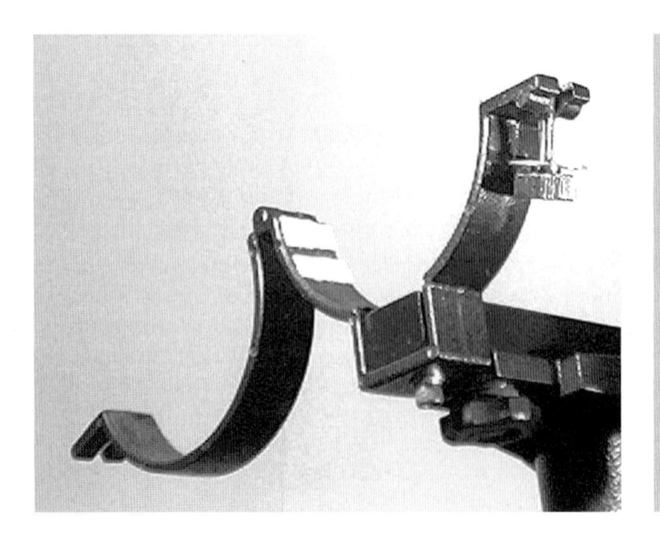

Figure 10.9. Felt pads positioned on the tube cradle rings to adjust optical tube angle during the adjustment operation in this section, shown here in white for clarity.

finder until its crosshairs also center on the mark. Make sure to balance the optical tube on the mount in the normal manner.

2. Tighten any motion locks. For mounts with no locks, secure the position of the head by blocking or taping it, or using another expedient that will fix it in place while centering. Mark the center (balance point) of the tube directly above the physical center of the cradle or mounting block, using a small indicator of tape or a colored-dot sticker.
3. Remove the tube from its mounting block or cradle, lift it up, then replace it securely and refasten, centering your tape indicator closely so that you know balance has not changed. Now, re-check for centering on the mark. In the average mounting, you will find that it has moved a bit (or even a lot) out of center. Try this a couple of times until you can reproduce the centering on your mark closely. You may find that tripod settling or some other factor other than the mount head is the cause.[12] Having remounted the tube, re-center your mark and reset all motion locks or the blocking-up arrangement.
4. Remove the tube a second time, and physically reverse it end-to-end on the mounting block or cradle, then tighten it in place, centering your tape mark so that fore-and-aft balance is preserved. Loosen the motion locks or remove the blocking, slew the tube 180 degrees and point to your mark, using the finder. Don't be surprised if the mark is off-center in altitude or even completely out of the field of view in the main tube. If it is, time spent on this procedure is vindicated. If not, you are fortunate and need proceed no further!
5. The finder field should be large enough to show the mark, and in which direction it has apparently moved. Loosen the tube and lift or depress the tail end while looking through the finder to determine how much change will be necessary, and mentally divide by half – remember, you need to split the difference. Steel-bladed "feeler" gauges of the type used for adjusting combustion engine valve and spark-plug clearances make relatively easy work of measuring a small gap, down to thousandths of an inch or micrometers. It will be easier in most cases to raise the cradle or block surface by placing accurate, permanent shims in place to raise "low" end of the tube in the cradle by exactly half the measured gap. This can be a permanent fix in most cases, given the small amount of movement required.
6. Gather appropriate shimming material. The thin felt or flannel buttons or strips made to protect table tops from scratches make excellent shims for a cradle mounting that uses felt padding at the points where it contacts the tube. Some trial and error will be required, since the material will compress slightly under pressure from the cradle ring. When altering a mounting block, brass sheet stock makes the best shimming material. Shim stock is available as a

---

[12] This "wandering" result is typical, and a reminder that the observer can benefit from developing a fastidious setup routine. It may seem extraordinarily picky, but with cradle or block-type mounts using knurled-wheel or nut fittings, I recommend memorizing roughly the number or turns or the angle of the fastener heads relative to some part of the tube once it is snugged down properly, repeating this routine every time the tube is remounted. Even a small torque-reading wrench is not out of place here, if the fasteners are of a type that can be or adapted for it.

single brass strip that tapers from paper-thin to substantial thickness. Steel blades from an inexpensive "feeler" gauge will work fine; with the advantage of knowing the shim's exact thickness. Such a shim can be placed *under* the padding on a cradle ring as well, since most padding can be carefully stripped off the ring and replaced after the gauge blade has been cemented in place with quick-setting epoxy or heavy-duty auto trim adhesive.

7. Measuring the required shim thickness in this way will put you close to the required offset. Perform another tube reversal, putting the pre-trimmed shimming material temporarily in place, and re-center your mark in the finder and main tube. Then, slew the tube several times, note any discrepancies, and adjust the shim thickness until you hit the mark with the tube mounted in either direction. Viola!

CHAPTER ELEVEN

Protection, Refinishing and Surface Treatments

Experienced observers realize, of course, that simple cosmetics don't compromise optical performance. John Dobson, Sidewalk Astronomer and inventor of the practical Newtonian configuration that bears his name, is justifiably proud of having made successful telescopes from what he calls "junk" parts. He emphasizes in talks that it is the viewing, not a prettified telescope, which conveys the true message of the night sky.

I still chuckle remembering the evening I helped out a trio of particularly klutzy observers who were setting up a homely 12-inch Dobsonian made of scrap wood held together with roofing nails and silicone sealer. There were even drips of yellow house paint on the spider vanes. The young woman in the group had taped a hefty book to the mirror box to counterbalance the contraption, but they needed a little more weight. I seem to remember contributing a brick that previously had served as an auto wheel chock. Nightfall, however, revealed pristine optics and a startlingly crisp view of the Cone Nebula. Amid mumbles in the background about hollow ribbed mirrors, dark matter and the Steady State Universe, I learned that these were graduate students from MIT who had concocted the scope from 1/20th-wave experimental optics in about 8 hours to check out a question I didn't have the math to comprehend. Lesson: You can't tell a book by its cover, in this case a bound volume of the *Astrophysical Journal.*

On another level, however, general cleanliness, surface finish and attractive coatings play a role in spreading the word about our pursuit. Although one endeavors to preserve the original finish on antique instruments, even here there comes a point where competent restoration is in order. Likewise, poor appearance in a modern telescope compromises value and gives a generally bad impression. In every field from winter sports to bicycle racing, the public have come to associate successful out-of-doors pursuits with well-designed and properly maintained equipment. Institutions often invite amateurs with telescopes for science education purposes. Likewise, we promote such events ourselves; it is fun and rewarding. The fact is that well-maintained and attractive scopes see far greater public use at such venues, which is, after all, the goal of being out there!

Care of Metal

Bare Metal Protection

Creeping corrosion of the bare metal parts of mounts and tubes is a constant aggravation. Special care is necessary in locations with high industrial pollution or marine environments with "salt air." Even chrome-plated tripod legs and fittings gather corrosion after prolonged contact with acidified dew. Use protectants on all such surfaces, and renew seasonally after cleaning with a solvent such as VM&P Naphtha or Acetone.

Oils: Non-acidic oils provide protection to areas such as steel counterweight shafts, focusing tubes, bolt heads, uncoated brass, exposed electrical connections and non-anodized aluminum surfaces. The following fit the bill:

- Key Oil, specified to be acid-free, is the lightweight petroleum-based product provided for wind instrument keys.
- Compass Oil, used in filling compass hemispheres, is a highly refined coal oil, available from marine equipment suppliers, and has similar cleaning and protective properties. It won't disturb most finishes, but works to remove stubborn adhesive residue.
- W-D 40© spray, the old mechanic's standby, is also quite useful for repelling water from dampened electrical connections and protecting interior spaces subject to corrosion; requires frequent renewal if exposed to weather.
- Petroleum jelly is a safe protectant and lubricant, but don't use on parts that will be handled – a thin coating provides minimal protection, while a thicker one transfers to any contacting surface.

All of these, used as a wipe-on agent, serve to remove acidic oils from hand contact, replacing them with a thin protective layer. Renew regularly. Apply them liberally to the surface so they penetrate under the edges of fittings and fasteners, and then remove surplus with a soft cloth, leaving only a thin coating.

Permanent Clear Coatings

Where permanent protective treatment is necessary for exposed metal, spray-applied clear methacrylate (acrylic), clear lacquer, or semigloss clear enamel make an effective weather coating. The ubiquitous polyurethane varnish of the woodworker is less useful in such applications, as it gives a glossy "dipped" appearance, and some brands crackle under temperature cycling when used on metal.

Anodized Parts: Anodizing forms such a tough coating that treatment is usually unnecessary for many years. Clean off oxidation or residue with lacquer thinner or acetone, using solvent-proof gloves, with adequate ventilation. Wipe a light coating of a metal protectant or oil on cleaned anodized surfaces to restore gloss. Waxing also works, although a wax stripper may be necessary to clean and renew the yellowed finish after time.

Caution: Spray-on silicone lubricants or protectants can also form a protective barrier, just be aware that they are inert, can't be easily removed, and will resist any future attempts at painting – the paint coating typically peeling away under light handling.

Removing Rust: Naval Jelly, sold in small plastic containers, works wonders in mild cases. Carefully clean off any rusted ferrous surfaces, using the measures described on the packaging. Steel wool in a fine grade (#000 to 0) is acceptably non-abrasive for most jobs. Follow by thorough rinsing and drying. While some jellies have a residual protectant effect, de-rusted ferrous metal is soon prone to oxidation; rinse thoroughly and protect immediately.

Exterior Metal Finishes

Protection and Touch-up The simplest protection is the kind one routinely gives a fine vehicle: washing, followed by careful waxing. Just remember, a vehicle has no surfaces as sensitive to abrasion or chemicals as your mirrors or lenses. Remove or protect them before cleaning. Mirrors, over-coated or not, are particularly sensitive to cleaning agents and abrasion, and should be removed, if practical, before any kind of extensive work is done.

If the scope is very grimy overall, this might be a good time to clean Newtonian, Classical Cassegrain, or other open-tube-mounted mirrors. (Chapter 7 covers mirror-cleaning methods.)

Pre-cleaning Removing Labels and Fixture-Mounting Adhesives: Scopes and equipment commonly lose original manufacturers' labels and warning stickers over time, leaving patches of petroleum-based adhesive behind to attract lint, dust and grime. They can be "rolled off" with the fingertips, but this takes time and leaves a messy smear. Equipment also acquires sticky adhesive patches from the removal of Velcro® patches and fixtures applied with mounting tape. The years have seen many ineffective, even damaging, recommendations on how to remove adhesive materials published. The following is based on museum conservation techniques:

1. Wash the area locally as described below under *General Cleaning*, and allow it to dry. This removes overlying water-soluble material that can block solvent action.
2. The safest, most effective solvent is pure-grade petroleum Naphtha, applied sparingly, as needed. Use the safe grade clearly marked "VM&P."

3. As always with any aromatic solvent, pre-check for effects by dotting an unobtrusive area with a cotton swab.[1]
4. *To remove adhesive remnants* saturate and press a slightly oversize patch of folded cotton fabric or paper towel against the area. Using latex gloves or a slip of plastic wrap to protect your fingers. The adhesive should dissolve within seconds, a few minutes at the longest if it is a few years old.
5. *To remove an intact label* or a stubborn patch of double-sided mounting tape, brush naphtha around the edges with a small watercolor or craft brush until the patch begins to loosen at a corner. Then slowly pull it off, flooding the released edge with fresh dabs from the brush. Wipe any residue away with cloth, tissues, or cotton swabs, using more naphtha if necessary, until the finish underneath is streak-free.

Tip: Patches of mounting tape hidden inaccessibly under fixtures can be worked loose by saturating a length of cotton string with naphtha solvent, and passing it carefully underneath the fixture (not against the equipment finish) with a sawing motion. Hold the wet string ends with gloves, foil, or plastic wrap. Resaturate the string as necessary. Once the fixture is removed, finish cleanup as described above.

Caution: Unlike most modern label and fastener adhesives, masking tape, duct tape, or fiberglass-reinforced tape left on for years can pose a problem. The adhesive can cross-link, becoming virtually insoluble. The fiberglass fibers remain embedded in the adhesive even when the paper substrate flakes off, posing a hazard to the eyes from casual handling. Even strong solvents such as methylene chloride may not be effective on such tapes, stripping the finish off, yet leaving the tape residue intact. Liquid citrus-concentrate cleaner can be useful. This may soften the paint underneath, however, since the process may require hours or days. One may try mechanical removal by scraping. Carefully pull the edge of a utility razor blade held perpendicular to the surface across the adhesive residue, trying not to slip and dig into the finish. *Wear safety goggles* to protect your eyes from flying hard particles of adhesive, fiberglass, or possible blade chips. One may need to follow up the procedure by retouching as described on page 186.

General Cleaning Commercial all-purpose or industrial cleaners with grease-cutting capabilities work better than plain soap or detergent for removing field grime. Use as directed for painted surfaces, and clean a small, unobtrusive test area first before proceeding with overall cleaning. Use soft cellulose-type sponges, clean cheesecloth, or clean, lint-free 100% cotton rag material for the purpose. No cleaner evaporates completely so play it safe, even with products that advertise they require “no rinsing.” Rinse several times with

[1] A few very substandard black finishes are naphtha-soluble.

filtered or distilled water. For best results, make sure *no* cleaner residue remains on the surface. Drying should be done with a clean chamois or toweling. Let the unit dry overnight if possible before waxing. Some moisture inevitably creeps underneath the equipment fasteners and fittings. Drawn out by the change in surface tension, it may run out and down, creating streaks and adding to the chore.

Bare Metal Protection Creeping corrosion of the bare metal parts of mounts and tubes is a common irritation. Even chrome-plated fittings spoil in prolonged contact with dew and mist, especially in urban or seashore environments. Use some sort of protectant on all such surfaces. Non-acidic oil or petroleum jelly is a very good surface protectant when renewed on a frequent basis. Non-acidic oil provides a measure of protection to areas such as counterweight shafts and focusing tubes where thicker lubricants tend to smear and foul clothing or accessories.

Naval jelly or other rust-removing compound will safely clean most rusted ferrous surfaces. Follow by thorough rinsing and drying. Protective wax can then be applied, just as with an automobile.

Wax Application Automobile wax or microcrystalline ("museum") wax are excellent for protecting most painted finishes, including enamel, epoxy and lacquer. These also work well for sealing breaks in the finish coating until doing retouch. Wipe a light coating of a metal protectant such as "W-D 40" on anodized surfaces to restore gloss after prolonged weather exposure.

Cautions: First, *never* wax a tube interior, lens cell, or the surfaces inside the dewcap or glare shield of a refractor objective. This will reduce the effectiveness of the non-reflective finish, and is difficult to keep away from the glass edge, any slips requiring fastidious cleaning. Second, pure silicones are virtually insoluble by normal cleaning methods. Unless you are prepared to sand or completely strip the finish from an area or component before any subsequent coating work, protectants containing silicone should in *no* case be allowed to encounter optical surfaces.

As with cleaning, use only cotton, cheesecloth, or the special wax applicators designed for fine auto or furniture finishes. Other materials may scratch or mar the surface you are trying to protect and enhance.

Avoid spray-on furniture wax products, for the reasons outlined above. On dark finishes, plain paste wax of the type used for wooden furniture is the easiest solution. This can pose problems on light-finished components like white optical tubes, however, because it tends to yellow over the long term. More refined waxes, such as carnauba waxes used on classic auto finishes, or microcrystalline waxes like the British Museum Wax used for book bindings and metal protection, give longer-lasting results. Generally, following manufacturers' directions for the wax product will give a neat protective finish that will resist dew and common stains, and allow easy cleaning for months, even under conditions of heavy dust and pollen.

Wax Stripping Removal using standard wax-removing compounds should present no problems. Follow product directions closely, however. If you strip wax, be sure to finish by a complete rinse with a strong detergent, followed by clear rinsing and towel or chamois drying.

Clear-Finish Coating Where extensive protection of a cleaned surface is necessary, clear methacrylate, acrylic, or lacquer applied from a spray can, spray gun, or airbrush makes an effective weather coating. Do this after thorough cleaning, and before any wax coating is applied, since the wax will prevent secure bonding or mix with the finish, creating hazy patches or rippling. Such a finish on bare metal obviates the need for lubricant or oil protection. Bear in mind that wear points will require occasional retouch where the coating wears off in normal use.

Before clear-coating a newly painted surface (see *Opaque Finishes*, below), allow sufficient time for the finish to cure, generally two weeks or longer, as directed on the packaging. Epoxy and powder-finishes are so tough that they should not require a clear-coat; epoxy will resist long-term bonding of standard clear-coatings.

Opaque Finishes on Metal Follow the pre-cleaning steps on page 182, and go to the procedures below. If you intend to refinish a component, be sure to remove all optics, accessories and mechanisms like focusers *before* the pre-cleaning phase. If this is impossible for some reason, thoroughly protect the mechanism during cleaning as described under *Masking*. Mask such areas before carrying out any filling or sanding steps as well.

Surface Preparation: When refinishing an existing enameled tube or other component in good condition in order to change color or add protection, simply take the gloss down first to promote coating adhesion, using a 400 or 600 grit wet-or-dry abrasive sheet.

On items with dent problems, start with automotive dent-filler of the two-part type, following directions. Sand the cured filler down flush with the surface, starting with 120-grit wet-or-dry, working down to 220-grit. Rinse and dry the surface with water to clear any slurry. Use wet-or-dry sheets *with* water, not dry, and work over absorbent newspaper or scrap cloth. After bringing all patched areas flush to the surface, use the same abrasive grits over the entire area, working down to 220 or 320 grit.

For filled spots or deep paint scratches and digs on a smooth, long surface like a tube assembly, use a conventional rubber sanding block or an improvised block of wood to confine abrasion to the high spots, keeping strokes smooth and consistent, dipping the block frequently in a water bowl to keep the surface wet and workable. This also cleans residue from the abrasive sheet, keeping the particles open.

Don't sand to bare metal unless you intend to use a metal primer before the finish coating. With a badly scratched item, a sandable automotive metal primer (gray) is a good filling medium, as it can be quickly worked down to the surface, yet remains in the scratches and surface digs. Use successive light coats followed by even sanding.

Clean all prepared and sanded surfaces with clean water and a sponge; dry, and follow up with lint-free cloth dampened in VM&P Naphtha. This will degrease and prepare the surface for coating.

Masking: Pieces of 4-mil polyethylene drop cloth or several layers of newspaper make an effective mask. Attach with painter's grade masking tape, and put a line or band of tape right to the edge of the area to be finished. Clear, self-adhesive shelf paper also makes a good mask for large areas where the underlying finish is sound. (Don't use this on "chancy" finishes, the tack of the paper may lift them when it is removed – stick to masking tape.)

Where tape or masking material overlaps the areas to refinish, trim back the edge with a fine-blade hobby knife, flush to the exact boundary and strip off the tape or mask. Use a straightedge if necessary to get a clean trim, being careful to cut through the tape while barely scoring the existing finish. Burnish down the edges of tape or mask adjacent to the area you will be refinishing, using a wooden brush handle or other smooth-surfaced tool. This will give a clean edge.

Finish coating: Use an exterior-grade paint of the lacquer, synthetic enamel, or epoxy type. Spray application is always better than a brush, unless you are an experienced brush-painter. Work in an open space with cross-ventilation or an exhaust fan, and protect nearby surfaces with drop cloths or newspaper. Follow the application directions on the container. Test on scrap material first, to determine how thick a coat you want, and how many coats you will need. Keep within the listed temperature and humidity restrictions of the coating material. Lacquer in particular is susceptible to cold, damp and hot weather conditions. You don't want to have to sand down a bumpy "orange peel" finish applied at the dew point, or find that your paint is drying as soon as it hits a hot surface and not leveling out.

Rust-Inhibiting Compounds Brush-on or spray-on rust inhibiting compounds that create an oxidation layer are effective and worth experimenting with as a primer undercoat for ferrous metal that will be painted, such as tripods, piers, or plinths for instance. Their characteristic dark coloration, however, will spoil the appearance of a polished metal surface.

General Touch-Up

Indelible Markers: Unless you feel lucky, spend a little time testing a finish you intend to touch up for a good match. Retouching nicks or bare spots with indelible markers or garden-variety flat paint can worsen the cosmetic effect. One technique is to use the point of the marker to dab, then immediately smooth the liquid out with a finger or a facial cosmetic sponge wedge, taking off the excess. Repeat the process, building up thin layers over the bare or scratched area. This will help reduce any apparent difference in gloss between the finishes. Finish up by smoothing the area with a cloth when the marker dye has dried, and use a little paste wax when thoroughly dry.

Paint Markers: Industrial retouch paint markers are useful for equipment, and come in a variety of surface glosses in many colors. Black, white and silver types will be the most useful for equipment. They are usually metal tubes with a shaking ball inside – shake thoroughly before using, and paint a test stripe or two to gauge the width of application. Use requires finesse, since the paint is actual enamel or lacquer dispensed onto a sponge or fiber tip by depressing the end of the barrel, and it is easy to apply too much. Keep rags and a non-aggressive solvent such as Naphtha nearby to clean up any goofs.

The gloss of a finish is an important factor, and no two surfaces are the same. Use a combination of the flat-finish and gloss types for most retouch work. Apply the gloss first over a bare spot or scratch. If it "disappears" on drying, that's fine. If it leaves too shiny a surface, try buffing it with a cloth to flatten it. If this is ineffective, then sequentially blend over the area with a small amount of flat retouch applied to a cosmetic sponge or small brush, followed by a little more gloss retouch, until the shine looks right.

India Ink: Another method that has been successful for touchup work on black surfaces is a combination of standard waterproof India artist's or drafting ink. This works particularly well for small chips on edges. You may enhance durability with a clear overcoat of nail enamel or semigloss spray (cover any optics thoroughly, of course, before using any spray for application).

Refilling Engraved Index Marks and Lettering

Scales on dials and rings serve a purpose. Beyond cosmetics, it's practical to be able to read a diopter or vernier scale, setting circle marks, or the maker's identifications on an ocular. The white filler material, usually enamel, commonly wears out of the markings with use. Where the surface is worn away, the channel is usually too shallow to hold filler. Where the engraving is relatively sharp and the surrounding surface smooth and intact (don't try this on heavily pebbled or grained surfaces), try the following:

Preparation: Petroleum-solvent cleaning, even with Naphtha, will generally worsen the condition by removing or smearing the filler, so clean the surface of oils with a moderate non-petroleum detergent solution such as Standard Solution B (See Appendix). Rinse with clear water, dry and follow with a very light coating of paste wax, applied with a soft cloth. Remove excess wax from the engraved marks with the points of wooden toothpicks. Don't let this stand more than half an hour, as it is there to assist in removing the residual after filling.

Filling: Use waterproof white drafting ink such as Pelikan™ brand, or one of the newer waterproof types of solvent-based office correcting fluids such as Liquid Paper Multi Fluid® or BIC all-purpose Wite-Out®. For a vintage look, use the fluid in ivory or buff.

Test by applying one of these to a sample marking, using a small brush. When dry, buff away any excess filler with the surface of a clean sheet of writing paper held flat. Don't use fabric or paper towel, since the idea is to skim over the engraving, cleaning only the flat surface around it. With a few trials, you will be able to remove the excess white, along with the thin coat of paste wax you applied. Repeat the process until the marks are clean and bright and the surrounding clean to the surface.

Clean filler out of any accidentally filled pits or scratches with the point of a toothpick and a cotton swab, and then follow up (on black surfaces) with a dab of India ink or ultra-fine-point black indelible marker.

Refill Black engraved markings on silver or yellow metals in a similar manner, using waterproof India ink, treating it as noted in the preceding section. Buff carefully to remove excess, disturbing the surrounding finish to the slightest degree possible.

Wood and Wood Finishes

Historic Wooden Tubes

Nowadays, wood finds its chief use in tripod legs, and in the mirror housings and carriages of Dobsonian-style reflectors and amateur-built instruments of various kinds. This has not always been the case. From the early days of the telescope right through the mid-19th century, wood was frequently used for the optical tubes of large commercial refracting and reflecting telescopes. For one thing, wooden tubes were not as hefty as those made with the heavy-gauge brass or rolled-and-riveted iron or steel available. Modern amateur workers have demonstrated that, up to a certain size, they can be just as rigid. Currently, in another shift away from all-metal tubes, some makers are experimenting with carbon-graphite composite, the lightest, strongest material yet developed. As with the labor-intense wooden tube, production costs are a principal restraining factor in this development.

The "classic" wooden tube was usually composed of long strips joined at the edges, much as a cooper would make a barrel out of staves. The classic long refractor tube, often conical or tapered at one or both ends, was usually built up in this fashion, planed smooth, and then scraped or sanded. Workers applied veneered hardwood, steamed to render it pliable. Depending on the locale, tubes were then finished with various techniques; most commonly, filler-stain was rubbed in and sealed with shellac, then varnished with multiple coats and waxed. The "golden" hue obtained often approximated the look of brass at first, but soon darkened with age.[2]

---

[2] The recently restored and remounted tube of the Great Refractor by Alvan Clark & Sons at Chicago's Adler Planetarium is a good example of an American 19th-century walnut-veneer production. The tube originally held the Clarks' 18.5-inch objective, largest in the world at the time.

Figure 11.1. Equatorial mounting constructed by J. Hannum and D. Radosevitch using laminated hardwoods. The tube is of similar construction. Photo: Loren Busch.

Richard Berry gives a good account of how to construct a simple square-section wooden refractor tube and alt-azimuth mounting in Make Your Own Telescope (see the Bibliography). Variations of tubes and mountings using modern birch plywood materials, even "butcher block" techniques, have been arrived at by contemporary ATMs.

Exterior Wooden Finishes

Some excellent Dobsonians and other wood-mounted instruments are put out in the field with slapdash finishes. This may have something to do with the division of crafts. Some skillful carpenters leave finishing to someone else, and never acquire the skill or appreciation for it. Still, it's a shame to spend time routing, rabbeting, dovetailing, or pegging joints for a rocker box, observing stand, or other piece of equipment, then skimp on the finishing steps. A basic method with variations suffices for general-purpose exterior wood finishing.

Surface Preparation

Pre-cleaning Bare Wood Prepare all woods before finishing, including "door skin" and birch or other hardwood laminates. The factory finish looks nice, but usually has surface defects, splintering, and oily areas from handling.

Pre-assemble: Put together all component units that will have a number of fasteners in them, using exterior grade yellow wood glue where the parts are permanently assembled, countersinking any finishing nails or wood screw heads. Fill the countersinks, BUT Match wood-filler color ahead of time:

Match Stain and Filler: Where you will be staining the wood, think ahead and prepare a color sample. Select a matching wood filler of the type you will be using to fill any countersunk fastener holes. Sand it down flush with the finish on the sample, and apply stain to test color compatibility, before beginning any finishing steps. You may find that darker or lighter filler gives better results. When you find the right match, make sure you have enough stain on hand for the entire job (see Staining, below).

Edge Trim: Identify any cut edges that will show after assembly and decide if you want to live with the unfinished appearance. If not, acquire an edge trim or molding. Use a matching veneered wood or a synthetic molding in a solid color, not a pseudo wood-grain. Some types wrap over the edge for splinter protection. In any case, choose one that has an inconspicuous fastening system or is compatible with a permanent adhesive such as epoxy.

After permanent assembly of components is complete, go over the surfaces and joints, carefully inspecting for adhesive drips or spatters that will resist the finish, showing as "white spots." Remove these mechanically, carefully back-scraping to the depth necessary with a flat-blade scraper or utility razor blade, working with the blade parallel to the surface and with the grain, to minimize disturbance.

Carefully cut away any beads of glue in corners or seams using the corner of a razor blade or knifepoint. Follow up by spot-cleaning such areas with wood alcohol or denatured alcohol on an absorbent cotton cloth. "De-grease" before sanding with cheesecloth, tack rags, or cotton material dampened with VM&P Naphtha. This will remove and keep from spreading finger oils and incidental oily materials, without raising the grain.

Sanding Start lightly with medium paper (150 grit or so) once overall, then quickly change to finer paper, finishing up with "fine" or "extra-fine" (300–400 grit). Always sand "with the grain." Cross-grain sanding will striate the surface trapping stain or finish. Blow off dust frequently to check progress, and make sure to scrape any additional glue spots that show up; re-sand to blend the texture.

Problem spots: Outside edges and corners – sand very lightly with 220 grit or finer paper to avoid splintering, finish up with fine grit paper on a sanding block. Do not round over the edges of solid wood by over-sanding – nothing looks more "hayseed" on an optical instrument. This is particularly important with softer woods like pine and poplar. Moreover, with fine veneers like Baltic birch or maple, over-sanding will quickly expose the core.

Tip: An exception is plywood where the edge shows in butt-jointed construction (unavoidable if you aren't using mitered joints) and threatens to splinter. Use rasping or belt sanding, create an even 1/4-round edge, and then fill any gaps with natural wood filler, building it up sufficiently. Take it down when cured with coarse- and medium-grit paper, finishing off with fine grit. Pay special attention to such edges when staining and sealing the wood before finishing.

Note: Use care with all coarse grit (60–100) paper when "taking down" projecting edges. A good alternative is a fine-tooth wood rasp, followed by medium grit paper. For all such operations, protect the adjoining wood surface with tape, and remove immediately to keep the adhesive from absorbing.

- Degrease again after sanding, using VM&P Naphtha
- Don't use water or detergent solutions on wood that will later be finished, except for necessary spot cleaning of water-soluble stains. Let such areas dry *thoroughly*, then re-sand to take down any grain.
- Wear solvent-resistant gloves while working with all solvents, and work in an open area with cross-ventilation or an exhaust fan.

Filling and Staining One option is to use a filler-stain-and-finish combination like Min Wax™ wood treatment. A single full coating may suffice. It gives an attractive semimatte or matte finish that looks like oiled wood, without the long process of oiling and curing. Additional coatings can be applied in the future, after first cleaning the surface. Alternately, finish if desired (after curing) with a proprietary coating, usually a synthetic varnish. Follow the manufacturer's package directions on such products.

Staining: This is for bare wood. Disassemble any removable parts for finishing; this will ensure that edges are treated and sealed. Use a standard oil-based wood stain, not varnish-stain or latex finish/stain. Test a sample piece of the wood to get the right color and consistency. You can mix shades or thin with mineral spirits to get the look you want. Be aware that most stains will darken by a shade or so when a finish is applied over them.

1. Mix colorant and filler material thoroughly from the bottom of the container and pour an amount of mixed stain into a glass or plastic disposable container large enough for stirring or with a top (for shaking). Stir or shake every few minutes while applying to keep the shade consistent.
2. Use a wide brush or foam paint pad for application. Wear gloves and old or protective clothing; this stuff won't come off in the wash. Work in an area with ample ventilation (outdoors if possible). Indoors, spread double or triple newspaper sheets over a plastic drop cloth under the work area and floor, extending farther than you think necessary.
3. Place scrap strips of wood under corners of the work-piece. Stain the entire surface quickly and efficiently at one time, starting from the top and working down, brushing against the grain at first to work the material in, then smoothing with the grain to avoid striping.
4. After the set-up period (see package directions), remove all excess using cotton rags, buffing lightly. Apply one to three coats, as necessary to achieve the tone you want.

When the stain is dry according to the manufacturer, which may take anywhere from 12 hours to a week or more, the finish coating can be started.

Sanding Sealer: This step can make the difference between a merely passable and a truly fine finish. Professional-grade sanding sealer has self-leveling proper-

Figure 11.2. A nicely finished all-wooden 22-inch Dobsonian mounting constructed by Hugo Jennings is seen here at the Riverside Telescope Makers Conference, 1999. The tube is so tall that a crane-mounted observing chair was constructed. Photo: Loren Busch.

ties and sands easily, so take advantage of the fact to prepare a smooth, lustrous finish for final coatings. Some brands have a proprietary solvent for cleanup, so read all directions carefully. For sanding, use fine grits only, 220 grit and finer, down to 320 or even 400. Apply one or more coats in the manner outlined below for the final finish.

Final Finishing Unless you have considerable experience with brushed coatings, use a spray finish to avoid brush marks. Just make sure that the type you use for a clear finish coat is compatible: that it won't "lift" the sanding sealer or stain underneath. Stick with one manufacturer's coating line, if possible, and follow package directions.

Tip: Get more of the product than you need to cover the project you are working on, setting aside a can for future touchups, or for finishing added parts that you want to match in gloss and durability.

Materials:

- Clear coating finish
- Wooden sticks, thin, to raise work pieces off of surface
- Wide and medium brushes for dusting and "grazing" off drips
- "Canned" air or an air compressor with air hose and nozzle
- Rags for cleaning
- Protective clothing: Rubber or fabric gloves, long-sleeved shirt, a cap to protect head from spray mist, wear eye goggles

- Plastic or fabric drop cloth or newspaper to protect surroundings
- Charcoal-cartridge type painter's vapor mask if working with spray coating indoors or in a confined area – in all cases, provide cross-ventilation or an strong exhaust fan for the room
- Proper solvent for the finish being applied, quart (liter) container or larger, depending upon the project
- Disposable containers for solvent-washing tools and brushes

Procedure: Wood finishing for exterior use, as with staining, is best done with disassembled parts. Many clear finishes state they provide single-coat coverage. Results will, however, usually be better with two or more quickly applied coats of finish, especially if you are using a type with a very short recoat time and a long curing time.

The toughest synthetic varnishes with highest weather and abrasion resistance tend to have long drying times due to resin content. With a finish that you know is going to need further smoothing, it is far more convenient to use a product such as spray lacquer or clear exterior-grade acrylic that dries to a sandable finish quickly. This allows convenient rubbing with fine abrasive paper (circa 400 grit) or fine (#0 or 00) steel wool between coats, to even out any remaining surface blemishes. This also avoids days-long finishing projects that can tie up home and shop space. You be the judge.

1. Lay out all the parts and assemblies, raised on thin sticks, over a bench or floor covering of plastic sheeting or newspaper. Immediately before applying the finish coating, go over all the surfaces with a clean, dry, wide brush, and follow with an air jet from a compressor or "canned air."
2. Start application immediately, to reduce dust or fibers drifting onto the surface. A fan-nozzle spray can or small air-compressor spray gun works best. Apply from your position and work away from you, advancing only half the width of the spray pattern with each stroke to blend the finish.
3. Don't use a continuous "looping" pattern; lift your finger from the valve or trigger at the end of each stroke, depress the valve and start back the other way.
4. Stop when a coat is complete, especially on vertical surfaces. With a slow-drying coating, you may be able to brush out any accidental drips or runs by light grazing with the tip of a medium-size bristle brush. Otherwise, let the piece dry thoroughly and correct the drip by sanding.
5. With pieces that require two-sides coating, apply one coat; let them dry thoroughly before turning them over to coat the other side. Turn over and finish the second side with multiple light coats, preventing running over the edges.
6. Turn a second time and apply the final multiple coatings to the first side
7. Do any between-coat sanding or rubbing with fine grit abrasives, and only after each coat is *thoroughly* dry, otherwise sludge will build up and slur the finish. Pay special attention to any bristly buildup along cut edges.
8. *If brush-coating*, be sure to check for loose bristles in the finish; lift and remove them with the point of a hobby knife while the finish is still wet, and then graze over the area with a bit of thinned finish on a small brush.

9. Do the final rubbing before the finish coat with fine steel wool, followed by a thorough cleaning with a soft brush and cloth, to remove any steel particles. Naphtha on a damp cloth will assist in this.
10. Allow curing time as directed on the packaging. Buffing with very fine steel wool will impart a nice semigloss finish. Otherwise, leave as-is, buffing with soft cloth to remove any dusty particles and remaining overspray.

A finish done in this manner will last for years and keep a good appearance. Apply lemon oil or paste wax seasonally to protect and keep moisture-resistant.

Renewing a Non-reflective Finish; Tubes and Accessories

Antireflective Coatings in General

The problem of scattered light plagued early telescopes; antireflective surfaces have been used since the earliest days. In the 17th and 18th centuries, tubes were often made from water-absorptive materials such as vellum or cardboard. These could be effectively coated with matte tempera or water-based paints. Later, deposited soot was favored for trapping reflected and scattered light from critical surfaces, particularly on brass. Makers wafted soot from candles or burning oil to settle inside the tube, blowing loose particles away before assembly.

Metalworkers up to the present day use chemical "pickling" processes, forming a durable black patina on brass for surfaces subject to light cleaning, and inside eyepiece barrels, lens cells and drawtubes. Mechanicians long used one relatively simple chemical method called "bronzing." The English optician Orford describes how bathing pre-heated brass parts in a copper-doped nitric acid solution can accomplish this. The worker then exposes the part to flame to oxidize the greenish deposits, achieving a durable, granular, matte-black oxide surface.[3] This is only one method. Texereau, for instance, gives a process for brass parts that his translator calls "blueing," using ammonium hydroxide and copper carbonate in solution.[4]

Some commercial makers regard interior coating as chiefly a cosmetic matter and coat only the most obvious surfaces. Although refractor tube interiors in particular can't be examined without disassembly or close inspection with a light, observers should make a practice of examining the inside of a prospective or newly acquired optical tube with a critical eye. The interiors of commercial optical tubes and accessories may have paint "skips," shiny drips, and chrome-plated fastener heads or bolt studs protruding into the light path. Light-baffle surfaces are sometimes uncoated or partially coated. Back reflection from these can result in irritating image ghosting on bright objects. In some optical systems,

---

[3] *Lens Work for Amateurs*, p. 95 – see the Bibliography.

[4] *How to Make a Telescope*, p. 257 – see the Bibliography.

significant reduction of image contrast can result. In other systems, complete interior coating makes little or no optical difference. In any case, a sloppy interior finish reveals poor artisanship.

Flat black coatings: Since anodizing processes can't yet produce an effective non-reflective surface on aluminum, optical tube and accessory interiors are usually finished with flat-black paint. Manufacturers employ a variety of spraying, dipping, or powder coating methods. The amateur can do all but the latter in the home shop, without any special equipment. The simplest coating is oil-based enamel, synthetic lacquer, or acrylic, usually incorporating a pigment (carbon black) with a flatting agent (wax or wax-like additive). Sometimes a standard "wrinkle" coat is used, sometimes spattered paint, silicate powder, ground charcoal, or fine grit is applied in a textured primer or mixed with the paint to further reduce surface gloss.

Given the relatively high proportion of flatteners and texturing agents necessary to achieve a very matte surface, such coatings are softer than the exterior finish, have minimal abrasion-resistance, and mar easily. A significant maintenance problem occurs when, as is commonly done, the maker uses the identical flat paint on exposed or potentially exposed surfaces such as focuser barrels, extension tubes, or dewcaps and glare shields. The surface can scratch or flake easily, and takes up stains readily. Oily deposits here darken with age, especially longstanding finger-smudges or lubricant smears, which are difficult to remove without disturbing the finish.

Cleaning

First, test a greasy or spotted matte finish with a swab moistened with VM&P Naphtha in a small, unobtrusive area. Several types of matte paint are not resistant to this solvent. If the swab picks up any pigment, stop there and go to a water-wash method. Parts that are Naphtha-resistant may be "dry cleaned" pretty thoroughly and oily contamination removed. Be sure to wear gloves and work in a well-ventilated space.

If you have freedom to immerse a metal part, such as a dew shield or a tube with all components removed, first test the resistance of the finish to a sample of household cleaning solution without additives (skin softeners or citrus fragrance, etc.). If it passes, simply soak the component as you would any painted item, following dilution directions for the product.

Clean long or large items outdoors. Use a hose to fill and rinse the washing container. A small, inexpensive plastic "kiddies" pool or wash trough works well. For tubes: professional paint stores carry long, disposable troughs for washing industrial-sized paint rollers. Rinse and dry in the shade if possible, not in sunlight.

Indoors, use a clean bathtub if it's large enough. Protect the surface from scratching with an inexpensive shower-curtain liner sheet, and rinse everything down well afterward.

Don't use abrasives of any kind, just a new sponge or cotton washcloth. Rubbing may reduce the matte qualities of a non-reflective finish. Clean gloss exterior coatings at the same time. You may be surprised at how much dirt comes off even a clean-looking scope tube.

Rinse in very warm water, and set the component on its end on an absorbent cloth to air-dry. Absorb beads of water with lint-free cloth or paper towel to avoid wire-edging stains that can occur if the water has minerals. Where you know this may happen, use filtered water for the final rinse.

Refinishing

Flat-black the interior before final exterior finishing, or carry it out as a remedial or refurbishing step on a finished tube. Either way, completely mask the exterior of the component. Use clear plastic sheeting, like drop cloth, as described for exterior finishes, and mask right to the edge of the protected area – right to the lip, for a tube. As with all spray coatings, apply finishes such as Krylon® Ultra-flat Black in a well ventilated area or outside.

Protection Tip: Flat black (and white) tend to hang in the air, drifting farther than other sprays, so drop-cover surfaces further away from the work than you might with a gloss enamel.

Finishing: The rougher the surface, the more stray light will be absorbed. Some workers sprinkle fine sawdust or crushed artist's charcoal in from the ends of an upright or tilted tube, right over a thick, wet, first spray coat, and immediately follow it up with a heavy second coat and another sprinkling of texturizer. This is messy but effective. Alternately, use a long-handled brush or a brush taped to a lath or molding stick to apply a thick layer of such a flat-black mixture as a first coat. Mix it up in a separate container before applying. Follow up with several spray coats, to firmly adhere any stray particles.

Caution: It only makes common sense to install any tube baffles *before* finishing with texturizer; otherwise, they will scrape the texture off on their way into the tube, creating more than one problem (see Chapter 13 on baffling).

To spray-finish a tube interior, textured or not, leave both ends open to avoid back-pressure, and aim the spray nozzle directly into the opening so the spray mist carries right through the tube. Spray for a few seconds, allow a small drying interval, and then rotate the tube $\frac{1}{4}$-turn and repeat.

Overspray is normally a problem in exterior finishes, but in flat-blacking, it is your friend. Try backing away farther than the normal spray distance, allowing the paint to partially dry, even drift onto the surface, creating a texture. With tubes, spray from both ends allowing the overspray to settle at the far end of the tube. Don't over-wet the surface, as this will reduce the texturing.

Apply two coats at minimum, preferably more. Drying time will be about double that for an exposed surface, so leave the tube open to the air at least overnight before mounting anything in it.

If you have used a texture-building additive, hold the tube upright after drying and tap the end sharply against the table to knock any detritus loose. Do this with both ends, and follow up with blasts of canned air or from an air compressor nozzle. Repeat this several times; otherwise you will inevitably be disassembling your tube to remove a fleck or two of the texturizer from the optics, an unnecessary task if the finishing had been followed up correctly.

CHAPTER TWELVE

Storage

Ideally, an individual or group will use and maintain all the instruments at their disposal, and the question of long-term storage will never come up. In reality, a host of factors operates against this. Factors such as lack of a suitable observing facility, changes of residence, questions of legal ownership, the need for extensive repair of accidental damage, etc. may entail retirement for an indefinite period. Also, the "out of sight, out of mind," syndrome must be addressed by both groups and individuals. No one wants to find out later that the storage area has had a rain leak all season, right above the telescope equipment. A friend's SCT stored in the back corner of a crowded shed suffered irreparable coating damage this way. Where possible, keep equipment and accessory containers away from outside walls, exposing all sides to view.

Shipping Boxes and Storage Containers

Telescopes almost universally ship from the factory in corrugated cardboard cartons. These give protection from mechanical shock and offer some water resistance, but are truly suitable only for semipermanent use. Adhesives used in box manufacture may attract and sustain silverfish or other insects. Moreover cardboard is acidic, pH <6.0 in most cases, and often includes an alum-rosin sizing material added to stiffen the product. The material, even when coated, is also water absorptive or *hygroscopic*, increasingly so over time as acidity breaks down the cellulose-size matrix. The combination readily outgases harmful compounds, especially when hot and/or damp. These substances will eventually attack glass surfaces or coatings through the action of sulfur dioxide and trioxide, sulfuric acid, peroxides and other harmful compounds in conjunction with atmospheric moisture and contamination. Unprotected metals stored in the cartons are highly susceptible to such attack as well.

Protective Plastic Coverings: Optics and mounts kept in cardboard boxes for more than a few weeks should first be thoroughly *dry*. As a permanent feature of storage, protect the entire optical tube assembly and any ferrous metal components with a mid-weight plastic bag. Mylar® is ideal, when available, but a polypropylene or polyethylene bag having a tight closure is an easier find. Vinyl or PVC film is not a good choice – impure grades may soften and stick to painted surfaces, or release decomposition volatiles such as plasticizers, which can deposit on optics. In general, avoid any plastic film with a strong odor or a gummy "feel" for use in long-term storage. The bag doesn't provide much physical protection, but will keep acidic vapor and stray moisture away from the optics and paint finish. Consider putting silica gel in the bags as well, and renewing it on a seasonal basis as long as items are stored (see *Silica Gel*, below).

Wooden Containers

Museums store their valuable scientific instruments in low-humidity, low sulfur environments. Some manufacturers offer storage cases made of tightly woven padded nylon, aluminum, sealed hardwood, low-acid wood, or inert plastic, which afford excellent protection.

Many observers construct their own cases for optical tubes. Home fabricators should be aware that most attractive hardwoods, especially oak, contain potentially harmful acidic volatiles that build up inside the closed case. Lebanon cedar or other aromatic woods, good for storing woolen fabrics due to insect-repellent properties, are not particularly kind to optics.

Seal the interior surfaces of all wooden cases well, preferably with a sanding sealer, followed by clear lacquer or color enamel. Allow such a finish to cure for several weeks before use. Epoxy is better for exterior use, since it usually requires a month or more for sufficient curing. Both alkyd and epoxy provide an almost moisture-proof seal.

Silica Gel and Humidity

Silica gel (silicon dioxide – SiO_2) is a non-toxic and relatively inexpensive hygroscopic (water-absorptive) material that is not only long acting, but re-usable. Universally utilized in factory packaging of optical and electronic goods (those little "Don't Eat Me" packets or canisters), it has recently also become widely available in bulk in the form of hard beads, crystals and even sheets specially produced for case enclosure.

The granular forms of silica gel usually come packaged in perforated containers. Most effective in a tightly sealed environment, the substance also keeps humidity down in less-than-airtight containers for short periods. Silica gel removes moisture that can condense onto surfaces during sharp temperature drops, as during cold-weather transport or air shipment. Over longer periods, the material keeps the relative humidity in containers down to a level that discourages oxidation and fungal growth.

Reusing Silica Gel: Gel saturated with moisture can be dried for re-use by removing it from the container and heating. Commercial packages carry instructions for this reactivation. Usually, spreading the granular material out on a metal cooking sheet and heating it at about 200 °F (94 °C) for an hour or two will dry it out completely, reactivating it for re-use. The melting point is above 2900 °F (1600 °C), so there is no danger of liquefying it in a normal home oven!

Caution: Silica gel granules are hard and abrasive. Packets of granules used around optics should be made of tightly woven fabric or a non-woven material such as spun bonded polyester like Tyvek™. Canisters must have tight, but air-permeable membranes over the ventilation holes to be effective but keep granules from sifting out. Don't brush or wipe any stray grains from optical surfaces; blow them off with an air jet, or pick up carefully with a cotton ball dampened in distilled water or rubbing alcohol.

CHAPTER THIRTEEN

... in keeping with the time-honored philosophy of amateur telescope making it has been judged desirable to retain in this volume some of the older methods, perhaps helpful to the isolated worker or hobbyist who likes to start, insofar as possible, from raw materials.

Allyn J. Thompson

Antihumidity Focuser Plugs

For years, observers have used the plastic canisters from 35-mm film for various applications, since they are a near exact fit in a 1.25-inch (31.8 mm) eyepiece holder. Make a handy desiccant canister from a film can pierced through the base to form a pattern of a dozen or so holes. Heat a large needle in a butane flame or stove burner to make clean holes. Take a few fresh silica-gel packets, stuff them into the film can, and use it instead of a standard plastic focuser plug for a refractor, "cat," large finder, or other closed-tube scope. The silica gel will slowly absorb humidity from inside the tube. The desiccant also keeps other accessories with 1.25-inch eyepiece receptacles dry in storage, including mirror diagonals, narrowband filter assemblies, CCD cameras and camera flip-mirror units. It can also work to a limited extent for a small Newtonian capped at both ends.

Ventilating a Newtonian Tube

Leaving ventilation openings below the cell at the bottom of a Newtonian tube is effective in hastening mirror cool down. Unfortunately, it also creates a "chimney effect." The passive nature of the convection leads to a shifting laminar structure of air layers within the tube that can persist for hours, robbing resolution from the image. As long as the tube is warmer than the surrounding air, currents lazily drift past the mirror and out the top of the tube, reversing direction when temperatures drop after cool-down, leading to cold air "pooling" around the mirror.

Nullify this by creating a deliberate, steady, airflow through the tube by mounting a small fan (small computer biscuit-fans are effective) behind the primary. It must be *isolated* from the tube and cell structure to avoid vibration. Creating a steady airflow homogenizes the lazy air layers, curing "local seeing" problems that originate within the telescope itself. The effect is dual, since the airflow also thermally stabilizes the primary.

Dr. Donald Parker was among the first of the fan-advocates to demonstrate proof of the curative effect in both film and CCD imaging. There are many references to methods for achieving this online and in the literature.

Baffling a Refractor Tube

This is something optical design manuals explain in detail but almost no mass-manufacturer of telescopes pays any real attention to. The fussy handwork simply adds greatly to production time. Even manufacturers of refractors, where the problem is most critical, often rely on simple, barely adequate light baffling, or "opaquing," as Rutten and Van Venrooij refer to it (see the Bibliography). Usually, there is just one metal baffle somewhere near the center of the tube, and the maker relies on the focuser tube to cut off the rest of the stray light. This is inadequate, unless the tube also includes the black flocked interior lining used by some quality makers.

The following project gives the basics on curing baffling problems for the typical refractor tube, primarily for home-built projects.[1]

Tools and Materials

- General tool selection – hex keys, jeweler's screwdrivers and large flat or cross-point screwdrivers for disassembling optical tube.
- Vernier caliper (helpful, but not necessary)
- Measuring tape

---

[1] Note: For your consumer protection, confine this project to tubes that are not factory sealed for Warranty purposes, and where the focusing tailpiece assembly and/or the refractor objective cell can be readily removed with simple hand tools. Apochromats and semi-Apochromats usually (not always) incorporate sufficient baffling, and any owner efforts may be (1) counterproductive and (2) void your warranty.

- Long straightedge (yard or meter stick)
- Utility knife
- Hack saw, metal file, abrasive paper and bench vise; for tubing cutting (optional)
- White banner paper or butcher paper longer than the telescope tube
- Matboard (description in text)
- Drawing or drafting compass
- Cotton rags
- Plastic bags or wrap for padding and protection of components
- Flat-black spray paint
- Jar lid larger than your focuser tube end
- Hardwood dowel, $\frac{1}{2}$-inch (12.5 mm), the length of telescope tube
- Silicone sealant, aquarium type
- Longest eyepiece you intend to use with the telescope
- Focuser extension tube (optional)

It is rather simple to control stray light in a refractor tube assembly with properly placed annular baffles. The solutions for a Newtonian can be rather complex, and have been recently published by Carlin.[2] With a refractor, however, the only figures you need are the optical tube dimensions, the focal length, and clear aperture of the objective, along with the outside diameter of the focuser tube. As with all work on optical tube assemblies, keeping the tube interior clean is critical.

The classic design method for determining baffle sizes mentions factors like "image scale," "fully illuminated field," and other fine points that are not practically relevant here, since you are not designing the instrument. We assume you will be modifying an existing tube that uses standard size eyepieces, and want baffles that clear the light path while effectively blocking stray reflections and off-axis light.

Focusing Tube Problems

We also assume here that the manufacturer has proportioned the focuser tube properly to admit the full light cone at infinity focus. You can roughly check this by focusing on infinity with your longest eyepiece, then pointing the tube at a bright wall. Look at the exit pupil from about a foot away. You should be able to see the thin, dark periphery of the objective cell (and the edges of any foil lens shims) symmetrically and very slightly intruding into the edge of the bright exit pupil circle. If there are no shims, place three tabs of masking tape at three spots on the perimeter of the cell lip (not touching the glass), so that they just intrude

[2] Nils Olof Carlin, *The Best of* Amateur Telescope Making Journal, vol. 2, p. 175 – see the Bibliography.

over the edge of the lens and take another look at the exit pupil. They should show up as small, dark notches.

Another check is to look with the bare eye through the system in daylight with the tube focused the same way. Remove the eyepiece and look through the focuser with your eye placed at the edge of the eyepiece holder. This is about where the field stop of the eyepiece would be. You should be able to see the periphery of the objective cell (and the edges of any foil lens shims) symmetrically and very slightly intruding into the edge of the field of vision. If you can't see these things, then the tube is too long. The baffling method described below includes a geometrical check, and a shop solution for this problem.

Working Method

Lay out a sheet of paper somewhat longer than your actual telescope tube on a hard surface; white butcher paper, computer or craft "banner" paper all work fine for the purpose. Use a #2 or other soft pencil and a long measuring stick to map out the actual light path of your instrument. If you don't know or want to double-check the manufacturer's stated focal length, take an eyepiece of your largest focuser diameter, mount it "straight through" (i.e. with no diagonal in place), and focus on infinity.[3]

Tip: If you don't have an extension tube that will allow this, now is the time to acquire one. Extension tubes are available for standard focuser sizes. Besides allowing you to view straight through, extension tubes expand the usefulness of a refractor with accessories such as Barlow lenses used for photography.

When focused on infinity, accurately measure the distance from the center of your objective lens to the field stop position of the eyepiece with a measuring tape or ruler. To do this, remove any dew shield and check the depth of the lens in the cell. Determine the center (roughly) by dividing the aperture by five; measure that distance down the tube from the front surface of the objective lens, marking this position on the outside of the cell by placing the edge of a tab of masking tape right at that point.

Note: You may obtain greater precision by removing the objective if possible, focusing the image of the Sun on an inflammable substance such as a ceramic tile, and measure the distance from the center of the objective to this focus.

If you don't use this focus-method, you need to take the field stop position of the eyepiece into account for your drawing. Calculate the field stop position of your eyepiece by removing the eyepiece, checking and measuring the distance of its field stop from the top. Measure down on the outside of the eyepiece holder and mark the point with a tape-tab edge. This corresponds to the focal plane at infinity focus. The distance between the tape tabs is your closely approximate focal length for astronomical purposes.

Now, make a diagram of the practical light cone from your objective: First draw a straight line (the optical axis) down the long dimension of the sheet of

---

[3] At night, use any celestial object or a miles-distant light source. In daylight, focus on something several miles away or the daytime Moon.

paper. Mark your measured focal length along this axis: draw a line at right angles to the axis at one end, the same length as the objective's clear aperture (split the difference on either side of the axis). Measure off the focal length, and draw a similar centered line at the far end, at right angles to the optical axis, the length of your focuser's aperture. Rule two more lines, top and bottom; run one from the top end-point of your "aperture" line, through the top end-point of the "focuser" line. Rule another through the bottom points, and extend the lines beyond their intersection, forming a long triangle with legs. This graphical representation shows the widest useable field you need to allow for your large eyepieces. The intersection is approximately where the image plane for 35-mm photography or a CCD or video chip will be.

Now, measure the inside diameter of the optical tube, and add a centered line outside the "aperture" line on your drawing. This measurement will be the outside diameter of your circular baffles. Assuming a cylindrical tube, duplicate this line at the "focuser" line position, and join the lines, forming a long rectangle to represent the inside dimensions of the tube. You can put baffles anywhere along the central axis. The aperture needed to clear the light cone at any point along the axis can be measured directly from a 1:1 drawing, or adjusted to whatever scale you used. The accompanying illustration gives the basic drawing layout and some typical baffle positions.

Mark the baffle positions you desire on the drawing, and transfer them to the tube by aligning it on the drawing, marking the optical tube exterior at the corresponding points with tape tabs. Draw a line through the center of each tab to pin down the exact position. Alternatively, measure the distances on the drawing and transfer the measurements to the tube.

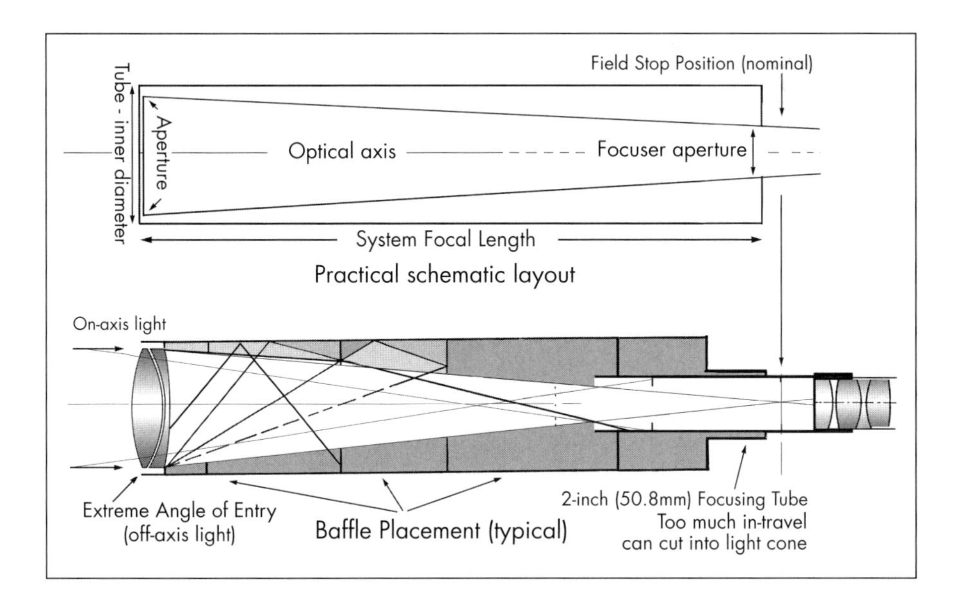

Figure 13.1. Baffle placement in a refractor (simplified).

Making the Baffles: After the locations and dimensions for the baffles are scaled out, a few hour's work with a piece of black, acid free picture-matting board, a compass and a sharp utility knife will do the trick. Draw the outer circular dimensions with a drafting or schoolroom compass, adding a bit to make sure they will be snug (you can always sand or trim them down slightly if they are too large), and then draw the inner aperture size with the compass point held in the same hole. Larger discs may require a carpenter's scribe with a pencil attachment.

To cut, use a scrap of clean board as a pad (not corrugated cardboard, since the corrugations will make the knife-line waver) and score a line lightly through the outer layer of board all around your circle to set the outline, then make several light cuts around the circumference until you are through the board. You may have to toss out a botched disc or two, but a little patience will yield good results.

Note: Having exact-sized discs precisely cut by a picture framer that has a circle-cutting machine will justify the nominal expense. For one thing, if you specify a bevel cut for the aperture (the outside edge should be a straight cut), you will automatically have "knife edge" baffles – these have virtually zero edge-thickness to catch and diffuse the light. Specifying black-core matboard in a black color will decrease the amount of finishing required. Finish the completed discs, including the knife-edge bevel edge, with several light coats of flat-black spray paint (see Materials List in the Appendix). When they are completed, lay them out in order from largest to smallest aperture.

Installation is a matter of sequentially pushing the discs into position from the front or back end of the tube. Use a hardwood dowel held against your optical tube, and mark the proper distance from the tube end to the tape tab marking each baffle-position with a pencil, then push each baffle to its correct position inside the tube. The dowel (or two) will serve well as pusher-sticks to adjust the baffles from either end of the tube.[4] If they are the correct size, friction will hold them in place until you have had a chance to sky-test the effects. If they are a bit under size, take up minor slack with thin strips of tape around the periphery of the discs.

When you permanently place each baffle, secure it with four or more small dabs of clear silicone aquarium adhesive around the periphery. In a long, restricted tube, apply the dabs by using a flashlight and a thin dowel for application. This flexible material will hold the baffles firmly in place, while slight tapping or a few hobby-knife cuts through the dabs will loosen them again if it is necessary to move or replace them.

Focuser Tube Light Cutoff: You may have detected a potential problem with the focuser tube, as mentioned above. The following procedure will determine the tube length to shoot for.

First, keeping the infinity-focus setting, remove the focuser and tailpiece from the optical tube and lay the unit on a 1:1-proportioned drawing.[5] Square it up so

---

[4] An existing baffle, if it's in the way, is usually cemented or very lightly soldered into position. Give a few sharp taps around its periphery, using a plastic hammer to strike the end of the 1/2-inch (12 mm) dowel you are using to mark and gauge the positions, then bend and pull the baffle out of the tube.

[5] If you are using a proportioned drawing, measure the distance from the field stop to the end of the focuser tube, divide it by the scale of your drawing, and draw a line this length forward on the optical axis from the field-stop position.

the eyepiece field stop is even with its position on the drawing. Using a short straightedge held against the end of the tube, mark the tube-end position accurately on the drawing, then measure the outside diameter of the focuser tube, not the inside, since a little clearance is a healthy thing.[6] Starting from the optical axis at this position, draw a centered line perpendicular to the axis matching the exterior dimension of the tube. With a 2-inch system, this will be around 2.125 inches or so. A 1.25-inch (31–.8 mm) or 0.965-inch (24.5 mm) focuser will be equivalently smaller.

If this line is contained within your light-cone boundaries, the focuser is potentially cutting off the light cone at infinity focus for imaging purposes. From the ends of this vertical line, draw two lines, parallel to the axis, back toward the field-stop position, to the point where they intersect the light cone lines. The length of this hypotenuse, plus a millimeter or so, is the amount you need to trim from the focusing tube.

Prepare the tube: Many focuser tubes are easily removed by unscrewing the eyepiece holder and racking them all the way in – usually there is an obvious setscrew in the focuser rack, a stop plate, or a removable stop that keeps this from happening in use. Determine this ahead of time, remove it, and remove the tube before marking it. If this is impractical, use the workaround below.

To determine your cut line, cut a piece of paper the length you have determined to trim off as a gauge (or use the depth marker on the vernier caliper). Mark this position at several points around the focuser tube (these units are usually sticky with grease; clean the end of the tube first with Acetone or VM&P Naphtha).[7] Place a band of masking tape around the tube as a cutting guide, right at your marks.

If the tube has to remain in the focuser, wrap the entire assembly in a plastic bag and tape it securely to avoid contaminating the focuser mechanism lubricant. Rack the focuser as far forward as possible, and clamp it in gently in the jaws of a vise padded with cloth (or hold the pre-wrapped focuser assembly firmly over the edge of your work surface, and go slowly). Make a straight, slow cut with a fine-tooth hacksaw. Clean up any edge burrs by scraping inside the tube with the tang of a file, or use a round file. Finish up with coarse and fine sandpapering to smooth the edge.

Blow any grit and filings out of the tube. Finish cleaning by pulling a clean, damp rag through the tube from one end to the other, and inspect the interior. Don't remove the plastic until this cleaning operation is finished. Clean a bare tube in the same way.

Note: Many focuser tubes are not well finished, and have bright ends to begin with. These should be flat-blacked while you have access to the unit, whether you

---

[6] With 2-inch systems, a 55 mm focal length 2-inch (50.8 mm) diameter eyepiece of circa 50° Apparent Field has its field stop at the inside edge of its chromed mounting sleeve. This is the largest true fully illuminated field theoretically possible in a 2-inch system – thus clearance for exactly 2 inches, plus a little, works fine for all practical purposes.

[7] Note: If the *focuser rack itself* extends over this area, place your marks short, i.e. to remove as much of the tube as you need to but without cutting through the rack – short of lengthening or replacing the outer tube, you will have to live with some light-cone restriction when using your longest oculars.

make a cut or not. Also, check the focusing tube for interior shininess. If there are paint-skips or shiny spots, hit the interior with a few short burst of flat-black spray. If you have cut the tube, you will need to flat-black the edge in any case. Avoid painting problems by spraying a small quantity of flat black into a disposable jar lid, and dip the finished end of the tube into the paint, holding it vertical long enough for the dip-coat to set up (for more on flat-blacking see Chapter 11).

With all the steps done, spot-finish any bright areas inside the focusing assembly, using flat black. Then remove the plastic protective wrap, and reassemble in reverse order, replacing any focuser stops or track setscrews and threading the eyepiece holder back on.

Making Finder and Eyepiece Cross-Hair Reticles

This work is not particularly difficult, but it requires patience. Oculars with a removable field stop are easiest but not necessary, since you can place cross-threads in nearly any style of positive eyepiece. Older Plössls, Modified Achromats and Kellners work well, and used spares for such projects are easy to acquire at swap tables or online trading venues.[8] Some newer small finders have etched reticles, as do the vintage 50-mm Polar Alignment Finders supplied with the original models of SCT. Otherwise, finders generally have integrated eyepieces and simple cross-hair systems that can be readily repaired or replaced.

General Disassembly – Finders and Eyepieces

Tools: A jewelers' screwdriver set (including flat and cross-points) and a selection of small hex keys in metric and SAE sizes will serve to disassemble most small instruments and eyepieces. A small optical spanner wrench, and flexible clamps are also invaluable. Use a vise to hold materials firmly while working. The illustration shows the use of a 1.25-inch (31.8 mm) flex clamp in removing a stubborn eyepiece barrel.

Setup: Work on a brightly lit, dust-free surface. A bench in front of a day-lit window is ideal. Alternately, use a gooseneck or cantilevered lamp to bring the light directly over the workspace. A computer mouse pad makes a good resilient surface for small work. It is better not to use a fabric pad over the bench, unless it is lint-free.

For holding cylindrical pieces, a small bench vise or clamp can be modified by adding sections of plastic or PVC water pipe sawed lengthwise to make good, flexible "clamshell" holder. Pad the halves with soft, lint-free cloth, adhere self-stick-

---

8 Note: As with all modification work, be aware that repairs or modifications by anyone other than an authorized technician void most manufacturer warranties. Confine initial project efforts to inexpensive out-of-warranty finders, and spare eyepieces.

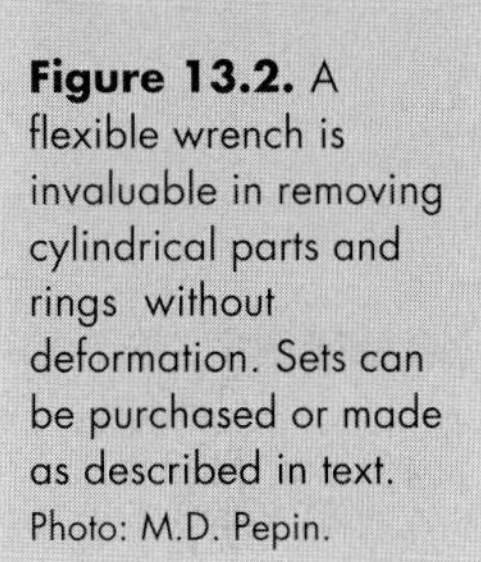

Figure 13.2. A flexible wrench is invaluable in removing cylindrical parts and rings without deformation. Sets can be purchased or made as described in text. Photo: M.D. Pepin.

ing felt padding, or cut sections of mouse pad (those old pads really are good for *something*). Temporarily affix the "clamshells" to the vise or clamp jaws with strong double-sided tape, or drill and tap the jaws for small bolts such as an 8–32 size.

Caution: *Don't* clamp an optical cell or eyepiece barrel tightly, or tap on it to loosen parts. This can easily chip or crack the lenses inside.

Removing the eyepiece barrel from an assembled unit, such as a finder, can be as simple as unscrewing it with your fingers. Alternatively, you may have to remove a threaded cap, rubber friction ring or eye shield to expose the setscrews, usually three, spaced 120° apart. The heads may be cross-point, slotted, or inset-hex.

Finder and binocular makers often conceal the eyepiece-retaining setscrews beneath removable rubber eyecups or in older models, under thread-on aluminum or Bakelite® eye-shields. Moreover, makers lacquer over screw heads in many optical fittings to lock them in place, or fill the holes with wax or rosin sealant. You can usually detect sealed locations by their uneven gloss, or by a slight depression in the painted or anodized surface. Use a small magnifier if necessary.

Remove sealed screws by first scraping any sealer out of the hole. Clean out sealed head slots with a fine-pointed knife blade until your screwdriver or hex key inserts cleanly. A drop of VM&P Naphtha, denatured alcohol, or acetone may ease the process, followed by cleanup with naphtha or water on a cotton swab to remove residue. Loosening the setscrews will usually allow you to unscrew or slide the eyepiece out.

When removing fasteners and small parts, put them immediately into a shallow container set well to the side to avoid inadvertent bumping. Keep track of setscrews (they are *tiny*) by sticking them to duct tape inside the container.

Checking the Eyepiece Handle the eyepiece carefully and don't touch the glass surfaces. Most thread-in types will have a coating of high-viscosity grease on the threads. If you are going to work on one, keep the grease clean and

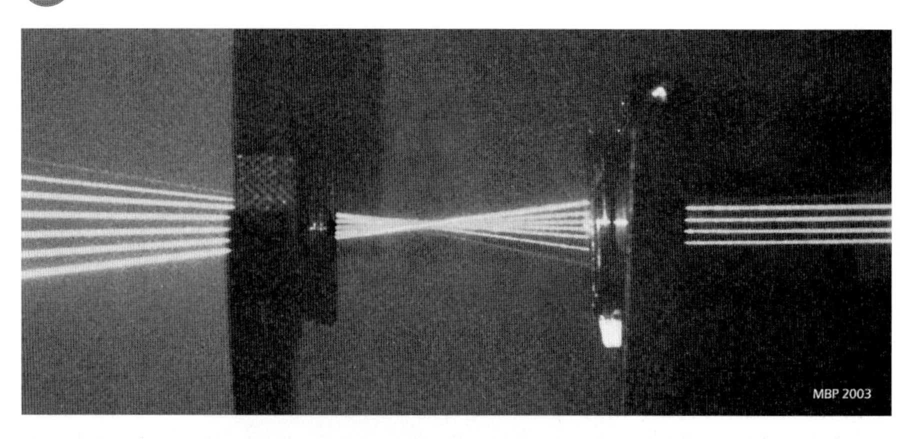

Figure 13.3. Laser tank close-up photograph of rays converging to focus between the lenses of a Huygenian eyepiece.

prevent transfer by immediately wrapping the sticky areas in metal foil or plastic film, securing with tape.

When you have removed the eyepiece from an instrument that has no existing cross-threads, check its type. Bend the tip of a toothpick or strip of paper over to form a right angle, and then look through the eyepiece while moving the object slowly closer until it comes into focus. This will generally be about a $\frac{1}{4}$-inch to an inch (6–25 mm) in front of the field lens. Be careful not to scratch the field lens, and don't check the focal point with your finger, as skin oils will transfer to the lens.

If your tester doesn't come into clear focus at any point, you probably have a Huygenian eyepiece with the focus between its lenses, as can be seen in the illustration. Makers mount threads in photometer and microscope Huygenian eyepieces, but the position is hard to reach, and the work of mounting and adjusting them goes beyond the scope of this manual. A Galilean eyepiece is negative, and looking through it will "minify" an object. By definition, crosshairs cannot be used with a Galilean system. In either case, use the opportunity to clean the lens and then reassemble!

If the eyepiece has a "positive" focus (beyond the field lens), but no field stop or ring, you can still fit cross-threads to the ocular by fabricating and mounting a "dummy" field stop ahead of the field lens, as described later.

Adding Eyepiece Cross-Hairs

Traditional Webs: The chief difference between a plain, small, spotter telescope and a finder is the set of eyepiece cross-hairs ("webs" in older sources) that allow the accurate centering of an object. Despite the name, natural hair generally makes a very poor cross-hair material. Even stiff horsehair distorts significantly with changes in relative humidity (remember the old science class hygrometer-dial experiment?). Hairs are too thick for most eyepiece purposes, in any case.

For these reasons, most instrument makers have used strong, moisture-resistant spider silk, often from the Black Widow spider, for ocular cross-hairs and micrometer webs. Specialty firms sold the material in carefully prepared rolls up through the mid-20th century. Roger Sinnott gives some details in "Cross Hairs from Spider Web" (*Sky & Telescope*, January 1987, p.97). Texereau also describes the process, recommending spider egg cases as a source.[9]

The durable etched glass reticles that have now replaced such webs for most purposes require precision photo-etching processes beyond the resources of the typical amateur. Scratching fine lines on glass or plastic is a quick-fix solution, but results have been generally unsatisfactory. Making a clear, consistent trace across the surface with a diamond scriber requires trial work, and a very steady hand. Furthermore, threads confer benefits over etched reticles for some purposes. It adds no glass to the optical system, with no surface intruded at the eyepiece focus to gather dust.

Threads Made from Liquid Cement

While spider webbing is no longer readily available, the astronomer who wishes to avoid working with arachnids can use clear household cement or styrene plastic to make cross-threads for most eyepieces. One can also mount cross-threads in older binocular eyepieces, since most of these are of the Kellner type with a field stop. The method can produce these in a range of thicknesses, in any pattern. A similar method works for replacing cross-hairs in riflescopes, the difference being the positioning near the midpoint of the scope barrel, rather than at the ocular lens.

Checking the eyepiece: If you intend to modify an instrument with an installed eyepiece, the first step is to check it for suitability. The common Kellner and Erfle type eyepieces found in most spotting and binocular instruments are the easiest to work with. These usually have an annular field stop in the correct position for mounting crosshairs. These are generally threaded or press-fitted into the bottom of the eyepiece head, or into the eyepiece-mounting barrel at the focal point.

Older Ramsden eyepieces may be used. With long focal lengths, focus usually falls far ahead of the eyepiece barrel itself, creating the need to extend the barrel. Adding one or more empty, threaded eyepiece filter cells or a barrel extender to reach the field-stop position works well.[10] The filter's threaded retaining ring can take the place of the adjustable field stop; do the mounting work with the retaining ring just seated at the top of its threads, then carefully thread the thin ring further into the cell to focus it after the threads are mounted.

---

[9] *How to Make a Telescope*, p. 265 – see the Bibliography.

[10] Several commercial sources now supply this item.

Working on the Eyepiece

The Field Stop and "Dummy" Field Stop

Note: Depending on your confidence and the ocular involved, some may wish to skip this next procedure and go directly to making a dummy field stop, below.

Procedure: Hold the eyepiece eye-end down under strong illumination and look beneath the field stop for the retaining ring that, usually holds the eyepiece's lens group in place. If there is a retaining ring, removing the field stop to attach cross-threads will generally be straightforward. The lens elements in some units, however (such as the 20-mm Kellner in the accompanying illustrations), are held in place by the field stop itself. These require more careful handling.

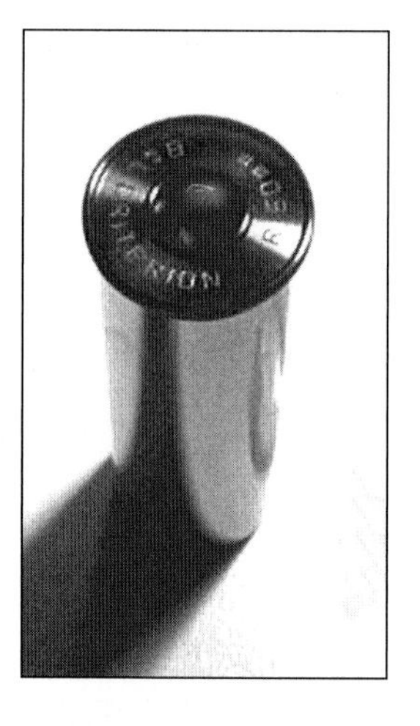

Figure 13.4. A vintage Ramsden 50-mm eyepiece by Bausch & Lomb; this type would require a barrel extension to attach crosshairs because the field lens is right at the bottom of the chromed barrel.

A threaded stop has its aperture set at the nominal focal point for a person with 20/20 vision. Usually the threads are lightly locked with dots of lacquer or glue. Most have slots that will fit your small optical spanner wrench.

Note: If you don't have an optical spanner, see Tool Tips under Optical Hand Tools in Appendix A.

In any case, do not remove the field stop all at once. If it is a threaded-in type, try to loosen it, and unscrew it by half to a full turn. You may have to use pressure to break the grip of the sealant. If the stop is pressure-fitted, loosen it slightly in the barrel with finger pressure or by gently pulling with the bent-over tip of a wire. Bend a plastic-coated paper clip into a puller that won't scratch the stop. Keeping the eye end of the eyepiece down, gently press upward on the eye lens with the end of a soft cotton swab or ball of tissue. If the eye lens shifts at all, keep the eye-end down while completely removing the field stop; then set the eyepiece carefully aside, covering it with a slip of filter paper to exclude dust. Do not turn it over or lenses and spacers may scatter all over the bench.

Caution: If the stop will not unscrew or slide out with moderate tool pressure, it is likely to bend or scratch, spoiling the edge. It is best not to attempt to mount cross-threads in a fully assembled eyepiece, as the adhesive may drip onto the field lens. Instead, fabricate a "dummy" stop as described below.

Cleaning the Eyepiece Interior The field lens of an older eyepiece often has specks or dust on it. Be sure the lens-retainer is tight before cleaning or displaced dust may find its way between the lenses. Blow any dust away with a gentle jet of air from a squeeze bulb or a pressurized air can. Then dust the field lens carefully with the tip of a clean sable watercolor or lens brush. Do not use any other type of brush, as synthetic fibers may scratch the lens. Lightly breathing on it and wiping the mist away with a cotton cleaning-swab may further clean the field lens.

Note: If the eyepiece barrel has sticky grease inside it, just blow any dust out and set it aside. Otherwise, you may have to completely disassemble the eyepiece to effectively remove any grease inadvertently transferred to the field lens.

Making a "Dummy" Field Stop A dummy stop works in any eyepiece if you would rather not disassemble it, if it has no field stop, or the field stop is not easily removable. This means you can safely put removable crosshairs on 2-inch (50.8 mm) ocular, for instance, making a gigantic finder scope. Unless the existing ocular has no field stop, the dummy needn't be opaque, in which case and you may use thin stock like a plasticized playing card, or even plastic cut from a milk jug or other container. For opacity, use thin-gauge brass or even aluminum (soda can bottoms work fine; are thin and easy to cut with small shears).

If you are covering an existing field stop, the center hole in the disc should be a little larger than the hole in the actual field stop: (a) the edges of the dummy won't constrict the view through the eyepiece and (b) you won't have to carefully smooth the inside edge, since it will be invisible. The dummy stop's outside diameter is not critical, as any size that fits inside the eyepiece barrel will work.

Note: If you are replacing defective wires or threads on a fixed field stop with a dummy, first hold the eyepiece with the eye lens facing up to keep material from falling into the eyepiece. Pull the old threads away with tweezers or small pliers by wiggling to fatigue the attachment points. Clean any residue away with a hobby blade to make a flat mounting area for the dummy stop, then wipe the surface clean to remove any flakes, following up with a few air blasts.

Making the Disc

With soft material: Trace outside and inside rings with a school or drafting compass, using two settings and the same center hole. Cut the center hole first with a fine-pointed hobby knife, then, cut out the disc with scissors or shears. Curved nail scissors are about right for making the curved cut in soft material like carton plastic or card.

Using thin metal stock, for an ocular without a field stop: Secure an oversize piece of the stock to a wooden block or plywood scrap using spray adhesive or contact cement. When the adhesive has set up, put the block in a table vise, face up. Trace the outside and inside rings, then drill out the center.

Make the edges of the aperture clean, as a ragged edge will show clearly in use. First, drill a small pilot hole centered on your compass hole, using a 1/16-inch (1.5 mm) drill bit, then slowly drill out to the proper size, using three or four increasingly larger bits, running each size of bit clear into the wood block. Remove from the block by soaking with a little acetone or Naphtha solvent, using persuasion with a thin knife blade if necessary. Be careful not to cockle the piece by prying.

Smooth the inside edge of the aperture with 220 grit paper wrapped on a small dowel, finishing up with 400. Wetted wet-or-dry paper is best for this work, since the wet grit will make slurry, and smooth the edge well. As the *last* step, carefully cut the disc from the work piece with metal shears or strong scissors. Make it a tiny bit smaller than the mounting surface, for leeway in centering. Test fit the dummy disc in the ocular. Complete by flat-blacking the outward (non-threaded) face of the disc. Then place threads on the dummy stop with the methods described for an actual stop below.

Mounting the Dummy Field Stop If the existing field stop is adjustable, thread it in or out so the edge of the stop is in sharp focus, then carefully test-fit the dummy stop. Carefully press the threaded side of the dummy against the existing stop without disturbing the threads. (Mounting the dummy stop with the cross-threads out may throw them out of focus). Before the adhesive sets, look into the eye end with the barrel pointed toward a bright surface to check the centering, and carefully adjust if necessary.

Styrene Threads for Low-Magnification Uses

Cross-threads made from drawn styrene plastic can replace the clumsy, thick wires found in many inexpensive finder scopes. This material is also useful for adding cross-threads or replacing defective threads in low power (4–8 ×) spotting scope eyepieces, and for modifying interchangeable eyepieces of long focal length (32 mm or more). These threads are too thick to be useful under higher magnifications.

The sticks of "sprue" or extrusion waste from plastic scale model kits are a good source of styrene. Alternately, employ a piece of a broken hard plastic toy or any styrene stick or rod. This stable material will make strong threads down to less than the diameter of human hair. The material is generally stable under normal humidity and temperature changes. Styrene threads also resist bending better than wire, and don't require coating to reduce reflections. You have the option to fit single or multiple cross-threads.

Caution: Do this project in an established work area. Work on a bench, not finished furniture. Put cardboard or newspaper under the work and on the floor to protect table and floor from minor drips or spills.

Tools:

- Pair of tweezers
- Small scissors
- A small sharp knife (such as X-Acto™ #11) or single-edge safety blade
- Masking tape
- Carpenter's white glue
- Matches, a butane lighter, or miniature butane torch
- Sprue rod from a styrene model kit, or a piece of scrap styrene cut into a stick shape, 4 to 5 inches long

Masking-tape an end of the plastic stock to a suitable piece of inflammable material such as a brick, a ceramic plate, or flat piece of scrap metal. Place securely on the worktable with most of the length of stock sticking over the edge. Heat the center of the stock intermittently with the match or lighter from the side. This will take only a few seconds of contact.

Take care not to over-heat the material; the plastic may ignite if the flame is kept on it too long: You can easily blow it out just like a candle, but the resulting threads will be charged with carbon particles and have a lumpy appearance under magnification.

When the stock sags and bends in the middle under its own weight, grasp the cool lower end (with tweezers!) and pull it steadily away from the table for 20 to 30 inches, drawing the hot material into a thread. If the thread breaks, start another one by pulling on the remaining hot material with the tweezers. As the stock cools, the drawing action will stop. To ensure a straight thread, keep it taut for a few seconds until it hardens. If both ends of the stock are still attached to the thread, let it hang straight by its own weight until cool.

The farther and faster you draw it, the thinner the thread will be. The center of the length will generally be the thinnest portion, and the diameter will be consistent for several inches. Cut the thread away from the stock with small scissors and lay it on the table. The thread is most easily handled with tweezers. Cut sections with a small knife or single-edge blade. Cut pieces slightly longer than the diameter of the field stop.

Mounting Styrene Threads: Make sure the field stop is clean. You may want to give it a thin coat of dead-flat black paint at this point. Be sure to cover any machined threads on the outer edge of the stop with masking tape before painting. If you are brushing it on, thin the paint slightly with the proper solvent. In any case, don't apply so thickly that it builds up around the aperture, causing a bumpy appearance at your edge-of-field. If this happens, clean the paint off with solvent and reapply.

If you have removed existing threads from the field stop, clean the old adhesive off; otherwise the new cross-threads may not lay flat on the stop. Mark the thread positions on the stop at 90° angles, or in whatever pattern you have chosen. Some manuals, such as Texereau, recommend filing a tiny channel with a triangular file at the attachment points; this does help with positioning, but may be impractical. In any case, for a basic cross-hair, mark positions for the new threads with a pencil at four equidistant points on the face. Use your jig (see Fig. 13.6) or a 360-degree protractor as a guide.

Place a tiny dot of white carpenter's glue at each point using a toothpick or knife tip as applicator. Carefully position each thread on the glue dot with the tweezers; make sure that it is straight. Place the threads so that they cross at the center. You will have a minute or so to shift them slightly before the glue sets. If the mounting is not satisfactory, just remove the threads, scrape off the glue, and start over.

After allowing a half hour or so for the glue to cure, trim off any protruding thread ends with the knife or razor blade.

Thin Threads for Interchangeable Eyepieces

Make very thin transparent threads from liquid plastic cement. These are useful for illuminated guiding Reticles or for non-illuminated cross-threads for short focal length, high power eyepieces. Any clear solvent-based resin cement, such as UHU™ Duco™ Cement, may be used. Threads of this material are dimensionally stable when cured, and will last a long time if protected from the elements.

The best eyepieces for the purpose are the traditional orthoscopic or Kellner types, often purchased for less than half the price of new configurations. Many have removable, adjustable field stops threaded into the barrel. In eyepieces with a focal length shorter than about 15 mm, the field stop will be a very small part, 1/4 inch (7 mm) or less in diameter. Even these small stops will accommodate usable threads. As when using thicker threads, one can fabricate and install a "dummy" stop instead of removing the actual stop.

Removing the Eyepiece Field Stop Most eyepieces have a chromed brass lower barrel threaded into the eyepiece head. The first step is to remove this barrel. It may take some coaxing, but the holdup is usually just a little antireflective paint or lacquer on the threads. If necessary, remove a stubborn barrel by wrapping it in felt and holding it gently in the table vise, using flexible tube clamp slip-joint pliers with the jaws wrapped in felt to hold and turn the head. Never clamp the head of the eyepiece in the vise, as this may deform it and shift the lenses. If necessary, also remove any knurled rings attached to the head. The field stop should now be accessible.

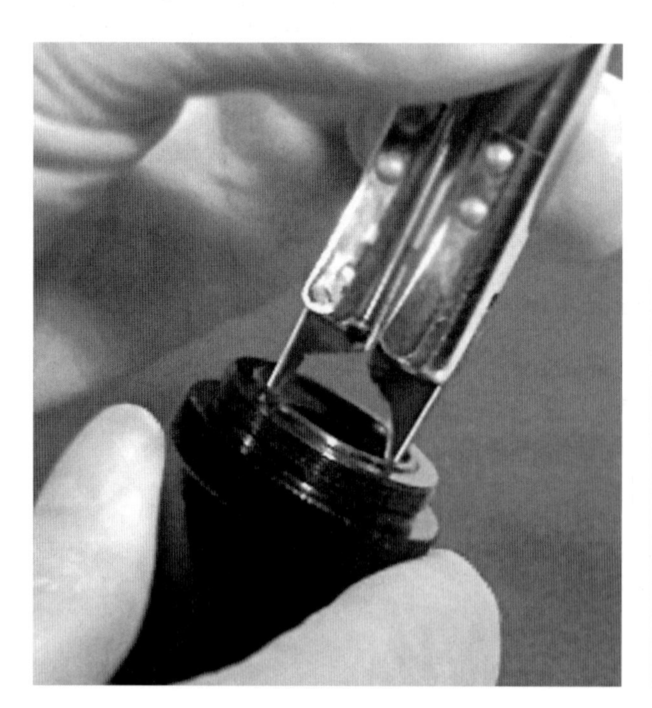

Figure 13.5. Using a small adjustable optical spanner wrench to loosen and remove the eyepiece field stop. Photo: M.D. Pepin.

Retaining Rings A thin, threaded retaining ring is the most common fastener holding lenses in cells. Opticians generally remove these by using a spanner wrench with parallel arms (see Chapter 8).

Note: If you don't have an optical spanner, see *Tool Tips* under *Optical Hand Tools* in Appendix A.

Whether using a homemade tool or a standard optical spanner wrench, it is a good idea to place a small disc cut from thin cardboard over the lens surface to protect it against slips while loosening the ring. You may also want to wrap the optical assembly in soft cloth to protect it from scratching. Apply even pressure until the ring begins to turn. You may also try unthreading the ring by rotating the eyepiece itself while holding the wrench steady. For additional stability, clamp the wrench in the table vise. Keep all movements steady, with no pressure directed toward the optical surface. If all else fails, a camera repair shop can usually remove such stubborn parts for a minimal charge.

Loosen the stop slightly with the eye end of the eyepiece pointing down, and then check for looseness. If the eye lens shifts or the eyepiece rattles when gently shaken, remove the field stop with the eyepiece in this upside-down position and then set it carefully aside.

Making and Mounting Cement Threads

One can carry out the process on any work surface, since no heating is involved. Protect finishes with cardboard or newspaper. Good lighting is essential, and it may be helpful to work over a dark surface such as black poster board to contrast with the gleaming thread. For tools, you will need a few toothpicks or small sticks, such as wooden matchsticks. Working under a stand magnifier or linen testing glass will make accurate placement easier.

Mark the edge of the field stop or dummy as for thick threads and place it squarely in front of you. A small jig made from artists mat board scrap or thin wood, marked with right angle lines (See illustration) can be most helpful for this.

Draw threads from a dab of the cement placed on a piece of scrap cardboard or a wooden spatula. One technique the author finds helpful is to dip the ends of two matchstick or toothpick applicators into a dab of fresh cement and then touch them together. As they are drawn apart, a fine thread will form, its thickness depending on the consistency of the cement. Drawing the thread out quickly (about a half-meter per second) may create uniform thickness and avoid sags.

Carry the thread over the stop and lower it carefully so that it touches both marks on the edge of the stop. Set the applicators down and straighten the thread by pushing gently down on it between the edge of the stop and the tabletop with toothpicks. Once arranged, set a small coin on the thread end to hold it in place. Secure the thread permanently with a dot of carpenter's glue on each end. A damp glue thread may appear to stick on its own, but will tend to fall off the stop in time, probably the first time you demonstrate your new cross-thread eyepiece to a friend.

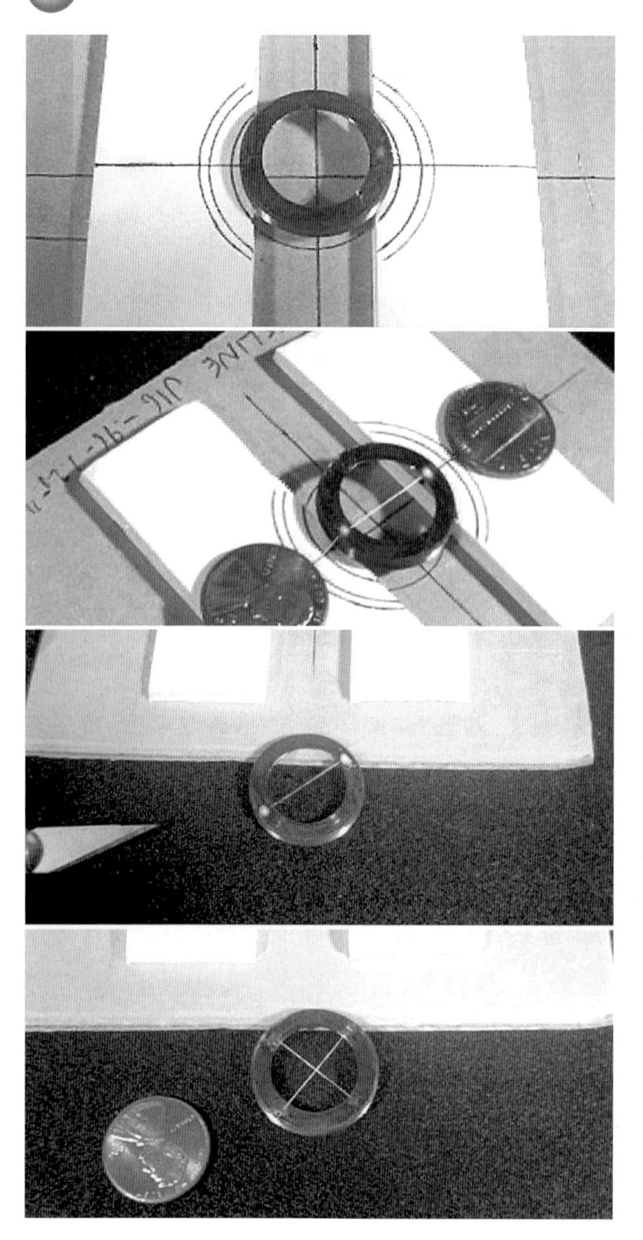

Figure 13.6. Stages of mounting cement threads on a field-stop. From top: Field stop on mounting jig; First thread glued in place, with coin weights to maintain tension; Single thread on stop; Finished cross hair reticle on stop.

Don't expect good results on the first try. A few minutes practice will allow you to judge the best way of drawing and placing the thread. This method can form threads almost as thin as spider silk. Since the placement is up to the maker, options include double or multiple cross-threads, offset cross-threads, even grids or patterns. As mentioned above, working under a stand magnifier on a jig, and checking the finished work with a glass will ensure results that are more accurate. After the glue has set well, trim off any protruding ends with manicure scissors or single-edge razor blade.

Figure 13.7. The field stop with cross hairs is threaded into the eyepiece barrel. Like spider webs, such "clear" threads appear opaque at focus, as imaged here with the eyepiece mounted in a small refractor.

Mounting and Reassembly You may wish to check and carefully clean the interior of the eyepiece before replacing the field stop. Use compressed air and a careful wipe with alcohol and a cotton or lens-paper swab, then recheck for particles.

Some eyepieces originally made with cross-threads or reticles have a helical focusing adjustment in the head to compensate for differences in vision. Standard telescope eyepieces, of course, do not have this feature. A field stop that holds the optical elements in place will not be adjustable, being preset for a user with nominal 20/20 vision. This is why the edge of the field-stop and the cross-threads in many finders' oculars are out of focus for near- or far-sighted users. In some interchangeable eyepieces, the field stop position is adjustable by screwing or pushing it in or out. This fact is never advertised, since optics makers seldom acknowledge such fine adjustment points. Take a close look to determine the possibility.

Tip: If you just hold the ocular up to the sky or toward a bright light, the shadow of the stop or crosshairs may give a blurred impression. Check for sharpness with the eyepiece placed in any small scope, or by looking through it at an evenly lit surface.

Carefully reassemble the field stop in the eyepiece without touching the threads, so that the threads are at the focal plane of the eyepiece, and in focus for your eye when you look through the eyepiece at a lit surface. If you wear glasses to observe, wear them to make this adjustment. This will customize the setting for your vision. If you have constructed a dummy field stop, adjust to focus on the existing field stop, and mount the dummy as described for "thick" threads.

Making an Eyepiece Occulting Bar Observers seeking to detect an object of interest near a brighter one – a faint companion star, planetary satellite, or a dim galaxy – often adjust the telescope so that the brighter object falls just outside the eyepiece field. The object becomes easier to detect in the darker background. Clark, Sidgwick, Roth and many journal articles have described variations on technique.

Using a Position Angle circle marked with degrees to determine the expected P.A. from an ephemeris or chart is almost a requirement for this technique, since it is a feat of coordination to search radially around the bright object without accidentally bumping it into the field, requiring the eye to readjust. Physical difficulties aside, it is particularly ineffective when using a telescope such as the SCT, Cassegrainian, or Newtonian, due to scattered light and diffraction from the secondary mirror, corrector plate backscatter, or spider vanes, which reduced contrast in the field with proximity to a bright object. In these cases, the simple occulting bar provides one solution.

Place an occulting bar on the field stop in the same manner as cross-threads, to block out the light from a bright object in the center of the field of view. The ocular itself can then be turned in its holder to scan radially around the object.

Regular cooking foil or aluminized polyester foil such as Baader solar-viewing foil should be painted flat black on both sides first, and then a strip of the correct width cleanly cut with a graphic arts knife (X-Acto™ #11 blade) or single-edge razor blade. Black graphic-arts film or developed photo film can also be used. Attach the bar and focus it in the same way as cross-threads. Use an isosceles triangle cut from a thin opaque film for blocking objects of different sizes and roughly measuring their diameters, as with a wedge micrometer.

Angular Width of Occulting Bar: You can determine the apparent angular size of the bar by comparison with lunar features of known diameter as listed in lunar guides. Alternately, use the method for determining apparent field in Appendix B; find the apparent angular width of the bar in the eyepiece using the wall-scale of known distance, then calculate the true angular width of the bar based on the magnification in the scope you are using.

Caution: **this is not a tool for solar viewing** unless used on a telescope that is already fitted with a proper full-aperture-blocking solar filter of Neutral Density of at least 5.0. A true occulting coronagraph has multiple baffles and discs optically aligned to occult the Sun without causing eye damage.

Removing and Cleaning the Finder Objective You may want to remove the finder or spotter objective for cleaning while installing cross-hairs in the eyepiece. Objectives up to 60-mm aperture are almost universally of the cemented achromat type, mounted in simple, one-piece cells with a single retaining ring. Objectives 80 mm and larger may be air-spaced, and one should check for the telltale metal foil shims, or a spacer ring, by shining a light into the front of the cell before attempting a quick removal. Refer to Chapter 8 for the disassembly and cleaning procedure.

Separating and Recementing Small Achromats

NOTE: There is no guarantee that the methods described here will be effective. Refer an instrument or accessory of high value or antique interest to a working optician or repair facility to determine if separation is wise or feasible. Pressure-cooking and various other drastic methods have been described in the literature, but the safety routines required, and the danger of damaging the lenses by overheating preclude recommending them here.

Separating a Small Cemented Achromat

Materials:

- Double-boiler pan
- Stove burner or hotplate with a heat control
- Cloth padding
- Immersible cooking thermometer (such as a candy thermometer)
- Tongs or large tweezers with padded tips (strong tape or heat-shrinking tubing)
- Dental pick, ceramic-working tool, or fine-point hobby knife, all metal
- Wooden craft sticks or chopsticks
- Protective clothing (thick rubber gloves, long-sleeved shirt, eye goggles, plastic or other waterproof apron)
- Solvents: Xylenes, acetone, isopropyl alcohol (isopropanol) 91% or better
- Glass eyedropper for solvent
- Canada balsam solution (as described in text)
- Wine bottle cork

Canada Balsam and Old-Style Lens Cements: The cement layer of achromats bonded with Canada balsam, rosin compounds, and many of the older synthetic cements will yellow with age after a few decades: sooner where there are impurities in the cement or under conditions of high temperature or long exposure to daylight. All of these accelerate the oxidation. In bad cases, the resin crackles or separates from the glass. This is usually curable by separating, cleaning, and recementing. However, it takes care and a deliberate working method.

Figure 13.8. The components of the achromat begin to slide apart in the bath at a temperature of 140° F (60° C). Photo: M.D. Pepin.

Figure 13.9. The cooled crown and flint components are separated and ready for cleaning.

Some achromats have taped or painted edges. Remove any tape or paint from the area around the lens seam, as it will hinder the process. With old, brittle tapes where the adhesive has become cross-linked, careful scraping may be necessary to bare the seam. With a blackened edge, try hot water and soap first (some optical firms have used ink for edge-blackening in the past). Otherwise, pure acetone usually works. This will not harm normal crown and flint optical glass or coatings. As noted several times in this book, do *not* use lacquer thinner for cleaning optical glass.

Important: As with separating an air spaced lens, first mark a bare portion of the glass clearly across both edges, making two non-parallel, slanted strokes with a soft writing pencil, so that you can identify the inner and outer faces of the convex element to properly reorient the two after separation. The figurer or polisher may also have compensated for slight "wedge" or other factors with local retouching, and such lenses need to be re-oriented to perform well. Pencil on bare glass will withstand the soaking process.

Water Bath Method: Lay out the achromat, padded tongs or tweezers, a clean pad to place the separated lenses on, and set the thermometer for just above pan depth, if it is the attachable candy-making type. Place the achromat in the pan on a suitable cloth pad (wash rag or cheesecloth) and cover it with plain tap water at room temperature, deep enough to cover the lens when held upright.

Start the heating process at the lowest possible flame or heat setting of a burner or range top. Slowly bring the water temperature up to simmering, but *never* to the boiling point. Home ranges or electric hotplates are not capable of the precise adjustments used for annealing when the achromat was joined in a dry oven, and the glass will not come up to temperature as quickly as the water will – the lag will be several minutes or more. "Slowly," therefore, requires constant watchful monitoring of the cooking thermometer inserted into the water pan. The aged balsam or synthetic cement will generally begin to lose its grip at around 140 °F (60 °C), balsam or rosin changing to a syrupy consistency or even disintegrating at around 200 °F (94 °C).

A smaller achromat, say 30 mm and under, doesn't require great care. Larger ones may require removing the pan from the heat at intervals to slow the heating process down, increasing the heat by only several degrees a minute until the water reaches a temperature of 140 °F (60 °C) by the thermometer. Hold the heat at this point for several minutes. Then tilt the achromat up carefully, holding it with the padded tweezers or tongs.

Heat up a dental tool (as shown in the illustration), or sharp hobby knife in a metal handle in the water first, then it can be drawn *very* lightly along the seam between the elements while holding the achromat in the bath. The action may open a channel for osmosis and accelerate the separation process. Do this underwater, holding the achromat with tongs. Don't pry, and don't actually insert a tool between the elements, as wedging action may crack or scratch the glass. Use wooden sticks to start the elements sliding apart if this seems necessary, but only when the grip of the cement has loosened. Use sideways pressure alone, and separation should proceed quickly from that point.

If this is not effective, increase the heat by stages as far as 200 °F (94 °C) but *don't bring the water to a boil.* With a stubborn achromat, remember that you can always stop at any point and try again later, but cool the bath down slowly first.[11] Do not pull the achromat or a lens fully out of the hot water bath while working – the thermal shock may crack it.

When the components completely separate, manipulate them to lie side-by-side on the cloth padding and start the cool-down process. Bring the water back to room temperature by lowering the heat in slow increments down to 120 °F (49 °C) or so, keeping the bath partially covered to slow the process. Remove the bath from the heat, bearing in mind that *the glass will retain heat much longer than the water will.* With a large lens (>50 mm), wait at least an hour after removal from the heat before removing lenses from the bath. No, this isn't paranoia – some outer meniscus lenses in particular are extremely thin and temperature-change sensitive.

When cool to room temperature, take the lenses carefully out with the tongs and lay them on a clean cotton pad. They will probably stick, but a few drops of the right solvent will remove them; this is a good way to test for it. Try purified xylenes (xylol or dimethylbenzene) first. It is usually effective, and specifically recommended by Johnson for optical balsam.[12]

Clean the residual cement from the fully cooled, dry lenses. Use old, soft, well-washed cotton rags or lens-cleaning paper, and never scrape: *dissolve.* Xylenes usually works, but turpentine or acetone or sometimes isopropyl alcohol (in the case of purified rosin cements) is also effective. Use gloves and good ventilation with these solvents; follow container directions. Again, do **not** use lacquer thinner or solvents of unknown composition, as they may fog or craze the glass.

---

[11] The heating process itself may "heal" crystallized balsam. If the" crackled" appearance diminishes in the bath before separation, try cooling the achromat down slowly as described, and check for this possible happy result.

[12] *Practical Optics*, p. 208 – see the Bibliography.

Recementing a Small Achromat

The reader is free to experiment, and synthetic lens cements are available from specialty houses. A small "hobby" quantity, along with full directions for competent use or assurances that the refractive index would be compatible, proved unavailable. Moreover, the newer two-part or light-cured cements are practically impossible to separate if you make a mistake.

Balsaming, on the other hand, is almost risk-free, and you can be certain that the refractive index is compatible with the separated elements you have on hand. The recommendation is to try balsaming the lenses in the traditional manner – optical quality will not suffer, and you can always separate, clean, and re-cement them with a synthetic if the opportunity arises.[13]

Pure Canada balsam solution is commonly available from microscope supply companies and specialized art supply outlets. Cover-glass balsam is a bit thin, but serviceable. The solutions sold as "medium balsam" by art supply outlets are generally better for optical purposes. You need only a small container (1 ounce or 30 ml), since an extremely thin layer is required to cement small lenses.

Working method: Trial-fit the crown and flint, checking the pencil marks you made before separation to identify the correct outer face of the convex lens. Turn this face downward, and set the lenses side-by-side. Both lenses must be scrupulously clean.

One standard method is to place the convex lens element on a felt pad on a cool warming plate, and heat it up until it is very warm to the touch, not burning hot. A home hotplate with a temperature control works well. (Again, make sure the proper convex surface faces up!)

1. Place a small amount of the liquid balsam on the center of the lens; roughly, $\frac{1}{4}$ teaspoon (12.5 ml) is more than enough for a 60-mm lens.
2. Pick up the concave element, check it one last time for dust, and lower it slowly and firmly onto the heated convex lens, making sure to orient your pencil marks.
3. Press lightly, straight down, then rotate slightly to seat the lens and drive out any bubbles. Use a bottle cork as a pressure tool, keeping it centered, as recommended by Twyman, Johnson, and Orford. The concave element will soon pick up warmth from the other lens, and the balsam should flow easily and bubble-free to the edges. Excess will flow onto the felt pad.
4. Press the elements steadily together until excess cement has been expelled, then turn the pair over and press lightly again to thoroughly warm the concave element, gently pressing with the cork to make the cement layer as thin as possible, but without creating areas of direct contact.
5. While still on the warming plate, clean the edges lightly with a clean rag barely dampened with turpentine.

[13] An alternative is to oil-space them. A drop or two of pure mineral oil between the lenses, followed by taping around the edge with a single layer of clear Scotch™ or similar tape, will secure the achromat for some time.

6. Test the lens alignment both visually and with a small straightedge around the periphery of the lens. Recheck the alignment marks and make any rotational adjustments.
7. While still hot, check the size of the seam between the glasses all around the periphery – it should look like a neat, even, slightly darker line rather than a gap. If it isn't, apply light pressure on the thicker side until the seam comes even all around.

Keep pressing, and after a few minutes, when a bubble-free fit has been obtained, carefully pick up the pair and look through them toward a day-lit window or bright wall, to determine that no detritus or tiny, unseen bubbles are trapped between the lenses. To adjust, return the unit to the warming plate, and then slide the lenses carefully apart while rotating to help break the seal. Clean them thoroughly in xylenes without abrasion, and repeat the process.

Now, older texts such as Orford and Twyman differ on the process. Orford recommends leaving the achromat on the warming plate to "set up" for at least an hour or so, Twyman recommends longer, and in an annealing oven. Either will cure the cement initially, and a few hours are necessary in any case for anything larger than an eyepiece lens.

Alternately, place the lens unit on a clean cloth on a baking sheet, then heat it up to a 'warm' setting (140 °F/60 °C), and bake for several hours. To cool, turn the oven off, but leave the door fully shut, letting the glass cool slowly back to room temperature for at least two hours. This will help relieve any strain between the glasses set up by the curing of the cement. Then, give the achromat a few days' time for the balsam to "cure" further, and clean the edges thoroughly. You may now remount it in its cell or eyepiece barrel. Optionally, blacken the edge with flat-black paint before mounting to eliminate a minor source of scattered light. Done!

APPENDIX A

Tools and Materials

Hand and Power Tools

Most removable parts and adjustable mechanical fittings in the optics world tighten down with inset-hex headed setscrews or slotted "grub" screws. Some manufacturers supply the minimum necessary toolkit for routine adjustments, such as adjusting the horizon altitude of the equatorial head. Others supply none at all. The adjustment of smaller assemblies often requires sizes not supplied, in any case. A set of metric "hex" keys or wrenches and a set of jewelers' screwdrivers are, therefore, indispensable for adjustment work on mounts, optical tube fittings, and accessories. Almost all manufacturers use the metric system, but there are enough exceptions even today to warrant a note here. Except for a few smaller sizes of hex key, you won't generally be able to make one set of tools work with equipment designed for the other.

For regular hexagonal nuts and hex-head bolts requiring wrenches (spanners), any standard metric or SAE set will work. British BSF (British Standard Fine), marked wrench sizes and listed nut and bolt sizes are identical to the American SAE, although earlier British producers generally avoided the intermediate sizes used in many US products, such as 11/32-inch, 13/16-inch, and so on.

In particular, acquire the proper hex-key set. Both the keys themselves and the sockets of inset-head screws can easily strip when using an undersized tool, and a tight hex wrench forced to work WILL break off and jam inside the socket head on the very afternoon when you need to quickly disassemble something for transport. Similarly, oversized spanners will slip, rounding off the flats on nuts and making them difficult to loosen again.

Hand Tools

- Screwdrivers, standard sized flat and cross-point ("Phillips" type), #1 through #4 point and larger as required
- Screwdrivers, "jewelers" or optician's set of six or eight with flat and cross-points (Phillips)
- Hex wrenches or "keys," – folding set or selection of "L" shape, double-ended; Metric sizes from 1.5 mm through 9 mm (and larger, depending on type of equipment used); "ballpoint" types are useful for angling into awkwardly-placed hex fasteners; SAE size equivalents: 1/16-inch through 3/8-inch

Optical Hand Tools

There are certain operations that benefit greatly from the following tools. These are available through professional tool outlets, optical suppliers, or in the used market (see illustrations in Chapter 8, on the disassembly and cleaning of small objectives).

Optical Spanner Wrenches These relatively expensive items are useful for general maintenance. In fact, they are the only tools that will allow you to remove most lenses or filters from their mountings without damaging the glass, the housing, or both. The adjustable optical spanner wrench is used to adjust tension or for disassembly. It generally has two orthogonally parallel shafts mounted to slide on parallel bars or a single trammel bar. The user separates the points on the shafts to match the spacing between holes or slots in the retaining rings that hold lenses and other items into their housings. Use the wrench with a firm grip and the optic steadily in hand; exert just enough pressure to "start" a tight ring. A small vise with thickly padded jaws makes an excellent holder.

Tool Tips: A suitable piece of thin steel sheet cut and filed to the right diameter is a good makeshift optical tool. Imported spring-steel machinist's and drafting rules are now avail-

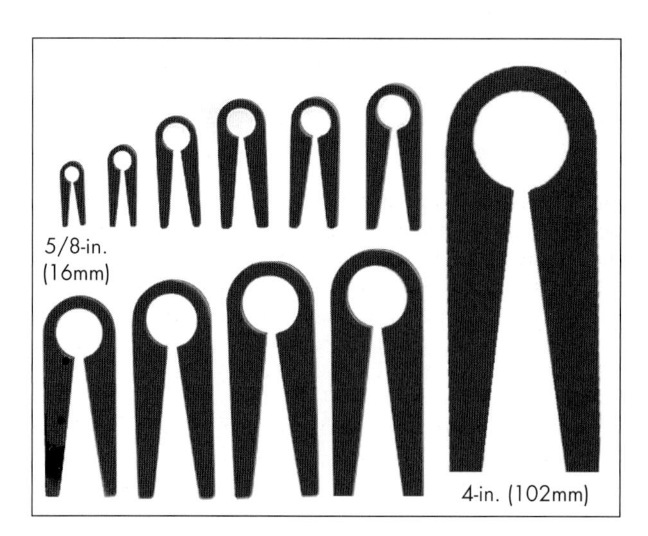

Figure A.1. Flexible tube and ring clamps hand-vises cut from sheet plastic or tempered hardboard will hold and remove threaded and fitted eyepiece and objective tubes without deformation.

able at the nominal cost of a few dollars at many supply outlets. The cost is usually less than a plain piece of the right gauge. A ruler of $\frac{1}{2}$-inch (12.5 mm) or 1-inch width (25.4 mm) cut and filed to different widths, will make half a dozen small flat-blade tools.

Flexible Tube and Ring Clamps

These are available in sets or individual sizes from several suppliers. The flexible plastic, composite, or spring-metal clamp simply fits over the assembly (see Fig A.1). The user squeezes the handle extensions while holding the unit by hand or in a vise, and exerts a turning motion to start the ring or barrel.

Fabricate them: Suitable materials are sheet stock of reinforced plastics like thick circuit-board material, acrylic sheet, or tempered hardboard. Thickness can range from 3/16-inch (6 mm), to thicker stock for larger assemblies. Make to any size by tracing (or drawing with a hand-compass) the required external diameter of the circular part, adding a fraction of an inch or a two-millimeter clearance. Trace out the handle extension desired. Cut with available tools. A hand coping saw or jeweler's saw works well. A band saw, if available, will make short work of it. A finish keeps the vises neat; black enamel as in the set shown works well (see illustration).

Internal Snap-Ring Pliers – Available with Various Pin Sizes

Small pliers with short pins mounted at right angles to the jaws. These remove standard internal snap rings with holes in their end lobes. One can also remove small, slightly recessed retaining rings and plates that have circular holes in their faces, over a wide range of spacing.

Power Tools

The worker who wants to complete operations more quickly will find several small power tools useful to speed up the shaping of parts. Among the types one might acquire:

Standard electric drill: Not really an option (there *are* a few necessities), it is the tool of all work, even if you happen to have a vintage-style hand drill around. Use bits as specified, depending on wood and metal hardness. Variable-speed/reversible types with rechargeable batteries are the most useful for field work. Acquire a set of tools for fastener removal and installation, and never leave for a star party without one—you may save your own or somebody else's night.

Miniature electric rotary tools: These, such as the Dremel® brand, are the traditional small power tools for the home hobby and craft worker. All makes have assorted bits and different-sized holding chucks for the power head; most have variable speed settings. Accessories include small abrasive cutoff wheels for metal and wood parts, drills, metal burrs for fine filing, shaping, routing, milling, and rasping; abrasive stones for plastics and different metals; miniature circular saw blades, and an assortment of buffing and finishing wheels; inexpensive assortments of bits and tools available, some with industrial diamond cutting edges for tough materials.

A flexible cable attachment is usually available, which allows suspending the motor from a stand while bringing the cutting head to bear on the work-piece from a convenient angle. Some makers also supply jigs and holders that facilitate use as a miniature grinder, drill press, or router.

Note: Follow manufacturer safety instructions. Always wear safety goggles, protective clothing, dust mask, and gloves while using the tool. Clean noxious or sharp granules and detritus from the work area frequently.

Abrasive Belt-and-Disc Combination Stand: These are available from many suppliers for bench-top use. The primary use is for light shaping and smoothing edges of wooden parts. When carefully used with the proper abrasive belts and discs (and eye goggles and a dust mask), works well for shaping a variety of stock materials. Metal-cutting abrasive belts are available, and you can use them instead of a mill or lathe to flatten metal stock surfaces, and to take small non-cylindrical metal, plastic, and composite material parts down to the proper dimensions for a project.

Wood and Metal Finishing Products

Wood Sealers and Stains

Oil-based stain: Traditional wood-shade ochre and carbon pigments and dyes, in an oil-solvent vehicle, require mixing before and during use; application by brushing and wiping off excess; long drying time; require clear finish coating for durability.

Sanding sealer: Lightweight filling and sandable "primer" coating for wood-finishing processes; use one or more coats with sanding under final coating for fine quality wood finishing and even surface texture.

Min-Wax™ wood finish or equivalent: Primary filling stain finish for wood protection, in tints for most wood finishes. Use as a final finish or coat with proprietary clear varnish after curing.

Coatings

Primers (metal): Spray-enamel type undercoats incorporating anti-oxidant chemicals or oxides; protect metal from corrosion; sand between primer coats to smooth surface, build up under smooth gloss or matte finish coat.

Primers (wood): Brush-on or spray wood primers are non-reactive, somewhat self-leveling undercoats; generally white, may be tinted; sand between coats with medium to fine carborundum paper to build up "grainless" finish.

Lacquers (generally formulated for metal or wood finish-coats over stain): Traditional nitrocellulose lacquer in spray cans or as a brushing type; clear in gloss, semigloss, and matte (flat) finishes. Makes excellent clear protective coating for all surfaces, metals in particular, including anodized aluminum; unavailable in some locales; non-nitrocellulose products may mimic the application process and appearance but are not "true lacquers."

Varnishes (wood): Slow drying resin-vehicle coatings; thin with mineral spirits; traditional coach or floor varnishes darken after time in most environments. The newer alkyd, polyurethane, and other synthetic and marine types recommended for exterior use; resist acid and alkali, but do not sand well between coats – make sure wood surface is as desired before applying; use care in application to avoid a thick, "dip coated" look

Enamels (metal and wood): Good quality enamel products are slowest drying, but provide lasting protection in exterior grades; brushing and spray enamels are available in clear and pigmented types of every color. Alkyd type is highly acid and alkali resistant. Wrinkle-finishes or "hammer finish" metallics in one- or two-part spray formulations mimic older finish types for restorations. Clear-coat type enamels protect metal-powder finishes with controllable gloss effects.

Tip: Automotive touchup acrylic enamels are good for exterior protection; hundreds of colors; can be matched for years; touch-up product available in small quantities and spray cans.

Krylon® and Plastic Paints: Flat and gloss types have a good reputation for resistance to weather; flat finishes retain matte appearance over time. Krylon Flat Black is extensively used for flat interior and baffle coatings. Contains a mix of several petroleum hydrocarbon solvents (with propane); ventilation is important

Note: A newer air-cured "absorptive polyurethane" paint, Aeroglaze® Z306 Flat Black, is among the flattest blacks in existence. Developed for outer space applications under rigorous conditions, it adheres to almost any substrate that has been properly primed; requires spray equipment for application; proprietary solvents for thinning and cleanup; 1-week curing time

Epoxy Paint (metal and wood): Synthetic resin paint available in many colors; dries quickly; the *one-part spray type must cure* several weeks before using inside containers or enclosures for optics; incompatible over other finishes. Follow container directions to the letter for good results.

Powder coating (metal): Professional 100% solids finishing option, requires special electrical heating equipment to melt powder and binder; now combined with anodized aluminum components as standard finish process on high-quality manufactured aluminum optical assemblies; typical finish is finely to coarsely textured; some powder coating facilities will service small customers for one-off finishing of items like piers, tubes, etc.

Wood Waxes, Oils and Touchup

Paste Furniture Wax (colorless): for final protective wax coating on finished wood. It may be used over thoroughly cured scratch-cover products and wood finishes of all types.

Wax touchup pencils: Fill small dents, digs, etc. in wood surface with matching tints; may be coated when cured.

Old English™ furniture polish: effective scratch cover and blending wax for light to dark wood finishes. Not highly protective, for application over existing finishes; requires re-application on a seasonal basis to maintain tone on soft woods.

Tung oil, Danish furniture oil finishes (wood only in general): slow-curing polymerizing natural-oil rubbing finishes for medium- and fine-grain hardwoods. Harden and darken with age; resist other finishes, a one-material, multicoat process, not for the quick weekend project!

Note: Generally avoid household waxes (for floors and counters, etc.) for this kind of finish work.

Coating Tools and Materials

Proper coating removal and application both require a modest assortment of tools and materials, but they must be of the highest quality.

Brushes and Applicators

Suffice it to say here that the cheapest grade of nylon and synthetic brush is hardly worth its weight in foam peanuts. To do a decent job of work, one should acquire at least a mid-grade of wood handled brush with a flagged synthetic sable bristle, or go all-out and acquire a few good flat brushes with the real thing. A good brush, well maintained, will literally last decades.

Wood and Metal Abrasives

For metal work, the "wet-or-dry types" are expensive but long lasting, and will do a better job than regular cabinet or wood-sanding paper on metal where primers or lacquer are in use. Use with water where possible.

- Coarse open-coat abrasive paper, 80 to 100 grit, for rust removal on ferrous metals and coarse sanding of wood and composites
- Medium/fine open-coat abrasive paper, 150 to 280 grit, for smoothing and burnishing ferrous metals and aluminum, between-coat finishing, and medium sanding of wood and composites
- Fine/very fine open-coat abrasive paper, 400 to 1000 grit, for polishing ferrous metals and aluminum and between-coat smoothing for finishing coats (lacquer or enamel), final sanding of wood and composites
- Emery cloth (red) charged with jeweler's rouge (iron oxide), for final surface polish rubbing of clear and matte finishes on ferrous metals, aluminum, and brass.
- Wet-or-dry (gray) paper is expensive, but best for fine finishing applications in conjunction with a water spray or wetted sanding block. It is durable, produces a finer surface, and has high aggressive power when used between finish coats.
- Carborundum paper (aluminum oxide), 80 to 300 grit for surfacing wood
- Steel wool in various grades from 000 (very fine) to 4 (coarse); useful for smoothing and de-burring surfaces; indispensable in the coarser grades for use with paint stripper after scraping (use with alkali-resistant gloves)

Useful Solvents and Coating Removers

Solvents

- Mineral spirits – Weak solvent, limited use for thinning and cleanup of oil-based enamel coatings, general cleanup. "Odorless" grades available for home use by solvent-sensitive people
- VM&P Naphtha (purified Naphthalene) – Petroleum hydrocarbon solvent for general light degreasing, and petroleum-based adhesive removal In finishing, use for surface preparation after sanding primers, and between-coat removal of sanding dust on metal or wood finishes; Will not disturb most cured coated surfaces, including enameled and varnished metal and wood, and *most* printing inks. Removes most adhesive labels; use with care around them to preserve adhesion.

- Toluene (Toluol) – Petroleum hydrocarbon solvent for general cleaning and mixing of paint coatings and cleanup, will remove accidental markings and surface stains of many types.
- Xylene (xylenes, dimethylbenzene) – Strong, odorous solvent for gums (including Canada balsam and pitch), oils, and waxes: Like toluene, removes permanent-type markers and most paints, also thins pure acrylic (methacrylate) resins. *Use with care* in open areas with good ventilation, will remove accidental markings and surface stains of many types.
- Denatured alcohol – **Toxic** but effective ethanol/solvent mixture, usually denatured with wood alcohol; may contain other irritant or toxic substances such as aircraft fuel; highly volatile, aggressive on varnish and enamel films.
- Wood alcohol (methanol, methyl alcohol) – Toxic solvent alcohol; works to clean up dried collodion; used primarily to thin shellac and to adulterate grain alcohol, sometimes substituted for isopropyl medical rubbing alcohol.
- Grain alcohol: (ethyl alcohol, ethanol) – Highly volatile, pharmacy-grade or "surgical" alcohol, available in double-distilled form (190-proof) in some areas; aggressive on varnish and enamel films; good for general cleaning and degreasing of optical parts, general cleaning of optics.
- Acetone (dimethyl ketone) – Good cleaner for coated and uncoated glass lenses. Use with care to avoid effects on adjacent materials, adhesives, and lens-edge blackening compounds. It is a good solvent for contact cements, adhesives, and polyester; generally use for cleanup of fiberglass and aluminum components, highly volatile, highly flammable.
- Lacquer thinner (proprietary varying mixture of toluene, isopropanol, and other solvents) – Cleans oxidation from aluminum components. **Avoid** all use on glass, due to etching or fogging effects on surface; thins lacquer, melts some tapes and plastics.

Paint Stripping Compounds

A variety of liquid, semipaste (slow drip), paste (little or no drip), and spray-on or "foaming" paint strippers are available from paint and home-improvement outlets. Instructions list the applicable coatings and surfaces, with practical instructions. Law requires instructions and warnings on containers, in packaging, and on product information.

Caution: Follow *all* label directions explicitly when using these generally harsh alkaline formulations, as painful skin and/or eye damage may result from exposure.

- Use in well-ventilated areas
- Use alkali- and chemical-resistant gloves – *minimally* latex painting gloves, changed often; rubber or neoprene is better.
- Wear protective clothing and/or shop apron
- Wear protective goggles in case of splashing
- Apply with recommended tools only
- Dispose of waste away from possible child contact

Application and a wait time is generally followed by scraping, a second application, then rubbing with coarse steel wool and small bristle-brush, followed by damp or solvent cleaning (wood) or a rinse (metal).

Mirror and Lens Cleaning Products

- Standard Solution A: Distilled water/detergent solution – 4 ml (1/8 teaspoon) liquid detergent (unscented, without skin-softening additives) or anionic surfactant cleaner – per gallon of distilled water
- Standard Solution B: Alcohol/detergent solution – 4 ml (1/8 teaspoon) liquid detergent (unscented, without skin-softening additives) or anionic surfactant detergent – per liter of isopropyl rubbing alcohol (91% or purer)
- Distilled water, for rinsing and making solutions
- Pure grain alcohol (ethyl alcohol): up to 150 proof – for specialized uses; a safer acetone-substitute cleaner for glass; not available in many areas
- Acetone – for specialized use in cleaning lenses, general cleanup and removal of petroleum based products

Advanced Mirror Cleaning – Materials Lists

- Tools required for mirror cell removal
- Pillow or folded blanket to support cell
- Sink or tub of adequate size
- Padding fabric to support mirror in bath
- Detergent
- Pure tap water, or filtered water or rainwater
- Distilled water to cover mirror and assembly, plus several gallons (+8–12 liters)
- Long-fiber surgical cotton, 1 package
- Blotter paper or cotton swabs
- Rubbing alcohol, isopropyl, 91% or stronger, 1 quart (1 liter)

Collodion Mirror Cleaning Process (also useful for lenses)

- USP Non-flexible Collodion (cellulose dissolved in ether and alcohol) Note: combustible material, extremely volatile, toxic when inhaled under USP standards, for use as strippable lens or mirror coating only. Observe all directions on packaging; see *Advanced Mirror Cleaning* for details.
- Cheesecloth (do *not* use this fabric for standard mirror or lens cleaning, not certified abrasion-free)
- Collodion solvent (proprietary, available from collodion sources) Useful for spills and general cleanup of utensils used with the collodion mixture: measuring cups, tweezers, etc. Wood alcohol will generally work also.

APPENDIX B

Practical Information and Formulas

Measuring the Apparent Field of an Eyepiece

Place the ocular being measured straight (use no diagonal) into a defocused small telescope in front of the eye, and turn it to face a brightly lit wall on which a measuring tape or stick has been mounted. (The defocused telescope has no optical function except to concentrate the light, sharpening the view of the field stop edge to increase accuracy.)

Facing the scale on the bright wall with both eyes open, the composite view obtained will be the image of the bright field circle seen through the eyepiece, superimposed over the scale. Center up the circle, and note the diameter of the field circle on the scale, along with the exact distance to the scale from your eye. Taking D as the diameter of the bright circle seen on the scale, and r as the distance from your eye to the scale apply the formula:

$120 \tan^{-1} [D/(2r)]$. This will accurately yield the *apparent field* of the eyepiece as used in any optical instrument.[1]

Light Gathering Power

The area of the primary optic is the key to light gathering power = πr^2. Taking the 7-mm full dark-adapted aperture of the eye as a base point, we can produce a proportional chart of light gathering relative to the eye:[2]

Eye (7 mm) = 1

[1] Thanks to Alan MacRobert for this useful method.

[2] Adapted from William J. Cook.

60 mm (2.36 in) = 74× the eye
80 mm (3.14 in) = 131×
90 mm (3.5 in) = 165×
101 mm (4.0 in) = 208×
114 mm (4.5 in) = 265×
127 mm (5.0 in) = 329×
152 mm (6.0 in) = 472×
178 mm (7.0 in) = 647×
203.2 mm (8.0 in) = 843×
235 mm (9.25 in) = 1127×
254 mm (10 in) = 1317×
279.4 mm (11 in) = 1593×
355.6 mm (14 in) = 2581×

Coma-Free Field of a Newtonian Reflector in Millimeters

(By Focal Ratio)

f/4.0 – 1.4 mm
f/4.5 – 2.0 mm
f/5.0 – 2.8 mm
f/6.0 – 4.8 mm
f/8.0 – 11 mm
f/9.0 – 17 mm
f/10.0 – 22 mm

List of Hints for Protecting Equipment

Avoid Weather Damage

When doing mobile observing, don't just drive all day to a remote location and immediately set up. Check weather reports and *query local people about unusual conditions* beforehand. Under a clear sky, one observer lost a good scope (and nearly his life) when a flash flood from a nearby thunderstorm roared down a "dry" gulch in which he was setting up to get out of the breeze. Any local could have told him ahead of time that he was crazy to do it. Other locations have regular seasonal afternoon deluges, localized wind anomalies such as "dust devils," etc. etc.

Keep It Dry

- Remove dew from tube and mount with a cloth or chamois during and after observing.
- Allow condensed moisture to air dry from *all* optical glass surfaces as soon as possible after observing.
- Let a dampened instrument and accessories dry completely before storing in airtight containers.

Figure B.1. Sketch by the author.

Guard Against Dust and Fungal Organisms

- Don't leave the end of the optical tube open to the air for long periods, and keep the open end downward or horizontal when uncovered.
- Turn the objective (and oculars in star diagonals) downwards when not actually observing to decrease atmospheric deposition of dust and mist.
- Cap up all optics when unused, but, again, *only* if they are dry. This includes anything that can be capped: objective lenses and focusers, star diagonals, eyepieces, etc. Use only caps made of latex rubber or inert plastics such as virgin polyethylene or styrene. Soft vinyl caps and unknown materials may outgas plasticizers.
- Dusting with an air-can and a medium-bristle brush will clean fungal spores out of crevices in equipment where they could bloom later,

Store in Cool, Dry Conditions Keep a portable telescope tube assembly in a protective case if possible. Many manufacturers supply these. Outlets listed in the Appendices offer generic protective bags and cases in many sizes.

- Use silica gel in all storage cases when possible.
- Don't store cases directly on or against bare concrete or cement block walls; this will saturate wood through transpiration and osmosis. Use plastic sheeting underneath as a moisture barrier.
- Don't store for long periods in unventilated exterior buildings, attics, or garages subject to extreme temperature cycling. If a telescope is left in an observatory or outbuilding unused for long periods, remove oculars and other sensitive components and cover the optical tube.

- Coverings: For a tight, dustproof fit use canvas, Gore-Tex, or any fabric that breathes. Humidity condensing under sealed plastic sheeting through diurnal temperature cycling will rapidly encourage fungus growth in warm periods, and rime deposition in winter conditions.

The Wratten Series Filters – A Partial Listing

Note: Filters listed in bold are commonly recommended for visual astronomy and are available in eyepiece barrel mountings. Listings include their overall transmission (T) in the visual luminance range for the most common thicknesses (~2–3 mm), and most commonly cited astronomical uses. The remaining partial selection is of filters (with their stated photographic uses) that have some contrast enhancing or other useful visual effects in the author's experience. Letters in parentheses are equivalent proprietary photographic designations. The information is adapted from various sources.

Wratten Filters and Characteristics

"Clear" Filters

UV (0) – colorless, absorbs UV
0 – clear, used for thickness compensation, i.e. in filter-wheel systems
1 – absorbs UV <360nm
1A Pale Pink – "clear" Skylight filter – widely used as protective window for camera lenses and at the Cassegrain focus in SCT and Maksutov systems. Slightly enhances contrast.

Yellow Range

2A Pale Yellow – absorbs UV to 405nm
2B Pale Yellow – absorbs UV to 390nm
4 Light Yellow – corrects outdoor scenes for Panchromatic (b/w) film
6 Light Yellow (K1) – partial correction for outdoors
8 Light Yellow (K2) – T= 83% – Similar to 11 and 15, general contrast enhancement on planetary surfaces, ring structures on Saturn
11 Yellow-green (X1) – T= 62% – surface detail on Moon, especially at full phases, cloud detail on gas planets, Venus
12 Deep Yellow – Uses similar to #15, for smaller apertures and "stacking"
13 Yellow-Green (X2) – corrects tungsten light for Type C panchromatic film
15 Deep Yellow (G) – T= 67% – lunar surface detail, especially at full phases, gas planet cloud features
16 Yellow-orange – blue absorption
18A – transmits UV and IR wavelengths only

Orange to Red Range

21 Orange – T= 46% – Enhances band and zone detail of gas planets, ring structure on Saturn, lunar surface detail
22 Deep orange/yellow-orange (mercury yellow) enhances contrast in blue for microscopy

23A Light red – T= 25% – like 25, weaker effects on aerial haze, good for smaller apertures and "stacking"
24 Red – two-color photography
25 Red (A) – T= 14% – Darkens sky background for daylight observations of Venus, Moon; Lunar and Martian surface detail; poles of gas planets
25A Red – Strongest b/w photographic contrast; absorbs UV and part of yellow, so-called night filter
26 Red – stereographic red
29 Deep red – tungsten projection of tri-color red in color separation

Magenta to Violet Range

32 Magenta – "minus green" filter
34A Violet-blue – color separation processes
47 Deep violet – (technically in "blue" range) – T= 13% – controls scintillation effects for Venus' phases and cusps; enhances "Blue Clearing" Martian phenomenon and Solar photosphere "plages" with ND filtration.

Blue and Blue-Green Range

38 Blue – red absorption
38A Dark Blue – T= 17% – Gas planet band and zone contrast; cometary gas tail enhancement
44 Light blue-green – "minus red" filter, two-color general viewing
47B Deep Blue – separation of color transparencies
50 Very dark blue – mercury violet light
80A Medium Blue – T= 30% – Lunar surface contrast; Martian polar caps and high atmospheric phenomena; polar caps, belts and features of gas planets; comet tail enhancement.
80B Blue - color correction for daylight film under 3400 K photographic lamps
82A Pale Blue – T= 73% – similar to 80A, useful in small apertures and for "stacking"; some control of secondary color in achromats

Green Range

52 Light green
53 Middle green
54 Very dark green
56 Light green – T= 53% – like 58, weaker effects, good for smaller apertures and "stacking"
57 Green – two-color photography
58 Green – T= 24% – Band and zone detail on gas planets, Martian soil contrast and polar cap "melt line" features, lunar detail under bright illumination
59A Very light green
64 Green – red absorption
66 Green – enhances contrast effects in microscopy and surgical photographs

Narrowband Range

70 Dark Red – IR photography – 676 nm
72B Dark Orange-yellow – 605 nm
73 Dark Yellow-green – 575 nm
74 Dark Green – mercury green – 539 nm
75 Dark Blue-green – 488 nm
76 Dark Violet (compound filter) – 449 nm

Neutral Density

96 Neutral Density – T= 9% with various transmissions up to ~83% depending on thickness – also characterized in the numerical ND range by T percentages, ND9, ND13, ND25, ND50, etc. "Moon" filter for bright lunar observation, Venus; use with objective Solar filters to reduce brightness.

Additional

RG630 – Energy Rejection Filter (Schott) – the "ERF" pre-filters infrared and ultraviolet radiation at the telescope aperture, used for Solar observation in the hydrogen-alpha wavelength with Solar prominence and narrow-band H-α etalon filters.

Scales of Seeing

Pickering Scale

William H. Pickering (1858-1938) of Harvard Observatory developed the scale, using observations with a 5-inch (130 mm) refractor.[3]

The *Association of Lunar and Planetary Observers* (A.L.P.O. – USA) employs this ascending scale, often broken into five "pairs," i.e. 1–2, 2–3, etc. to match the *descending* Roman numeral Antoniadi scale (see below).

- 1 – *Very Poor*: star image 2× the diameter of the 3rd diffraction ring
- 2 – *Very Poor*: star image occasionally 2× the diameter of the 3rd diffraction ring
- 3 – *Poor to Very Poor*: star image about the same diameter as the 3rd diffraction ring and brighter at the center
- 4 – *Poor*: Airy disc often visible, with "arcs" of the diffraction rings sometimes seen
- 5 – *Fair*: Airy disc always visible, arcs frequently seen
- 6 – *Fair to Good*: Airy disc always visible, short arcs constantly seen
- 7 – *Good*: Airy disc sometimes sharply defined, diffraction rings seen as long arcs or complete circles
- 8 – *Good to Excellent*: Airy disc always sharply defined. Diffraction rings seen as long arcs or complete circles, but always in motion
- 9 – *Excellent*: inner diffraction ring stationary, outer rings occasionally stationary
- 10 – *Excellent to Perfect*: the complete diffraction pattern is stationary

Antoniadi Scale

A simplified scale developed by E.M. Antoniadi (1870–1943). A Greek astronomer who long worked in France, Antoniadi was Director of the Mars Observing Section of the

[3] This is among the most definitive of the "subjective" scales. The descriptions of variations in the star image are rather easily discriminated by the keen observer using a star within the medium range of limiting magnitude for the instrument.

British Astronomical Society from 1896–1916. This is the scale used by most amateur observers in Europe and the U.K.

- I – Perfect seeing without a quiver
- II – Slight undulations, moments of calm lasting several seconds
- III – Moderate seeing with larger air tremors
- IV – Poor seeing, constant troublesome undulations
- V – Very bad seeing, even a rough sketch impossible

Relative Brightness

The Relative Brightness between systems of a system in use can be evaluated by applying this simple formula, where *B* is the relative brightness, *D* is the aperture of the system, and *FL* the focal length of the system:

$B = (D/FL)^2$

a. For an f/5.6 system of 100 D and 560 FL:
D×D / FL×FL = 10,000 / 313,600 = .0319

b. For an f8 system of 100 D and 800 FL:
D×D / FL×FL =10,000 / 640,000 = .0156

Given identical illumination, objective or mirror "a" delivers twice the brightness (total illuminance) of "b" to the same measured area at the focal plane. This demonstrates why adjacent photographic f/stops require twice the exposure time.

APPENDIX C

Short-Lists; Equipment Suppliers

Astronomical Equipment and Specialties

USA

Anacortes Telescope and Wild Bird
9973 Padilla Heights Road
Anacortes, WA 98221
USA
Phone: (360) 588-9000 Fax: (360) 588-9100
Currently online at: www.buytelescopes.com

Astro-Physics, Inc
11250 Forest Hills Road
Rockford, IL 61115
USA
Phone: 815-282-1513
Currently online at: www.astrophysics.com

Orion Telescopes and Binoculars
P.O. Box 1815
Santa Cruz, CA 95061-1815
USA
Phone: 800-676-1343
Currently online at: www.telescope.com

Tele Vue Optics, Inc.
32 Elkay Dr.
Chester, NY 10918
Phone: 845 469 4551
Currently online at: www.televue.com

Canada

Sky Instruments
Optics for ATMs
MPO Box 2037
Vancouver, BC V6B 3R6
Canada
Phone: 604-270-2813

Kendrick Astro Instruments
2920 Dundas Street West
Toronto, ONT M6P 1Y8
Canada
Phone: 416-762-7946
Currently online at: www.kendrick-ai.com

UK

Telescope House
63 Farringdon Road
London EC1M 3JB
Phone 020 7405 2156 (6 lines)
Currently online at: www.telescopehouse.co.uk

Beacon Hill Telescopes
Hope Cottage
112 Mill Road
Cleethorpes
Nth East Lincs
DN35 8JD
England
Phone: +44 (0)1472 692 959
Currently online at: www.beaconhilltelescopes.mcmail.com

Belgium

Lichtenknecker Optics NV
Kuringersteenweg 44
3500 Hasselt
Belgium
Phone: International: +32 11 253052
Currently online at: www.lo.be/lo/en/index.htm

Germany

Baader Planetarium, GmbH
Zur Sternwarte
82291 Mammendorf
Germany
Phone: (08145) 88-02
Currently online at: www.baader-planetarium.de

Vehrenberg KG
Meerbuscher Straße 64-78
40670 Meerbusch-Osterath
Germany
Phone: (02159) 5203-21
Currently online at: www.vehrenberg.de

Equipment and Tools

Most telescope companies advertise in the astronomy journals and are readily identifiable by brand name. Here are a few low-profile specialty firms that offer astronomically useful tools and supplies:

Edmund Industrial Optics
101 East Gloucester Pike
Barrington, NJ 08007-1380
USA
Phone: (800) 363-1992, Fax: (856) 573-6295
Currently online at: www.edmundoptics.com

S.K. Grimes Company
153 Hamlet Ave, 5th floor
(P.O. Box 1724)
Woonsocket, RI 02895
Phone:(401) 762-0857
Currently online at: www.skgrimes.com

Newport Corporation
1791 Deere Avenue
Irvine, CA 92606
Phone: (949) 863-3144
Currently online at: www.newport.com

Rolyn Optics Company
706 Arrowgrand Circle
Covina, California 97122
USA
Phone: 800-207-1312 or 626-915-5707
Currently online at: www.rolyn.com/

APPENDIX D

Popular Astronomical Journals

Astronomy
Kalmbach Publishing
21027 Crossroads Circle
P.O. Box 1612
Waukesha, WI 53187-1612
USA
Phone: 262-796-8776

Astronomy Now
PO Box 175
Tonbridge
Kent TN10 4QX
United Kingdom
Phone: +44 1732 367542

Astronomie Heute
Spektrum der Wissenschaft
Boschstrasse 12
D-69469 Weinheim
Germany
Phone: (06201) 6061-50

Ciel et Espace
l'Association française d'astronomie
17, Rue Émile Deutsche de la Meurthe
75014 Paris
France
Phone: 33 (1).45.89.81.44

Mercury
Astronomical Society of the Pacific
390 Ashton Avenue
San Francisco, CA 94112
Phone: 415-337-1100

SkyNews
Box 10
Yarker, Ontario K0K 3N0
Canada
Phone: 866-759-0005

Sky & Telescope
Sky Publishing Corp.
49 Bay State Road
Cambridge, MA 02138-1200
USA
Phone: 617-864-7360

Sterne und Weltraum
Max-Planck-Institut für Astronomie
Königstuhl 17
D-69117 Heidelberg
Germany
E-mail: quetz@mpia.de

Zvezdotchet
Zvezdotchet Magazine
P.O.Box 2
Moscow
Russia 119002

Bibliography

Amateur Telescope Making Journal, *The Best of* (Willmann-Bell, Richmond, 2003) 2 Vols., ed. William J. Cook

Barlow, BV, *The Astronomical Telescope* (Springer-Verlag, London and New York, 1975)

Beck, Rainer, et al., *Solar Astronomy Handbook* (Willmann-Bell, Richmond, 1995)

Bell, Louis, *The Telescope* (McGraw-Hill, New York, 1922)

Brown, Sam, *All About Telescopes* (Edmund Scientific, Barrington, N.J., 1993, 10th ed.)

Byrne, John, *Catalogue of Astronomical & Terrestrial Telescopes, Achromatic Object Glass* (Edward J. Brady, New York, 1874)

D'Auria, Tippy, and Menard, Vic, *Perspectives on Collimation* (Tectron, Sarasota, 1988)

Danjon, André & Couder, André, *Lunettes et Télescopes* (Paris, Editions de la Revue d'Optique Théoretique et Instrumentale, 1935)

Dickinson, Terence, and Dyer, Alan, *The Backyard Astronomer' Guide* (Firefly, Buffalo, 1994)

Fillmore, Warren I, *Construction of a Maksutov Telescope* (Sky Publishing, Cambridge, 1961).

Gingerich, Owen, ed., *Astrophysics and Twentieth Century Astronomy to 1950* (Cambridge University Press, 1984)

Glover, Thomas J, *Pocket Ref* (Sequoia, Littleton, Colorado, 1997, 2nd ed.)

Harrington, Philip S, *Touring the Universe Through Binoculars* (John Wiley, New York, 1990)

Ingalls, Albert G, (Ed.), *Amateur Telescope Making* (Scientific American, New York, 1953) 3 Vols.

Johnson, BK, *Practical Optics* (Hatton Press, London, 1946, 2nd ed.)

King, Henry C, *The History of the Telescope* (Sky Publishing, Cambridge, Mass., 1955)

König, Albert, and Köhler, Horst, *Die Fernrohre und Entfernungsmesser* (Springer-Verlag, Berlin, 1959, 3rd ed.)

Kriege, David, and Berry, Richard, *The Dobsonian Telescope: A Practical Manual for Building Large Aperture Telescopes* (Willmann-Bell, Richmond, 1997)
Lockyer, J Norman, *Stargazing, Past and Present,* (R Clay, Sons, and Taylor, London, 1883)
Manly, Peter, *The 20-cm Schmidt–Cassegrain Telescope* (Cambridge University Press, 1994)
Manly, Peter, *Unusual Telescopes* (Cambridge University Press, 1991)
Molyneux, Samuel, FRS, *Dioptrica Nova, A Treatise of Dioptricks In Two Parts* (Benjamin Tooke, London, 1692)
Moore, Patrick, *Exploring the Night Sky with Binoculars* (Cambridge University Press, 2000, 4th ed.)
Muirden, James, *The Amateur Astronomer's Handbook* (Harper & Row, New York, 1983, 3rd ed.)
Myers, Robert A (ed.) *Encyclopedia of Astronomy and Astrophysics* (Academic Press, New York, 1989)
Orford, Henry, *Lens Work for Amateurs* (Sir Isaac Pitman & Sons, London, 1944, 5th ed.)
Ottway, WR & Co., Ltd., *Catalogue of Telescopes, Supplies &c.* (Ealing, London, *ca.* 1945)
Paul, Henry E, *Binoculars and All Purpose Telescopes* (Amphoto, Garden City, New York, 1980)
Paul, Henry E, *Telescopes for Skygazing* (Amphoto, Garden City, New York, 1967, 2nd ed.)
Pepin, M Barlow, *The Emergence of the Telescope* (T Tauri Productions, Duncanville, Texas, 2002)
Roth, Günter D, (ed.), *Astronomy: a Handbook* (Springer-Verlag, New York, 1975) trans. & rev. Arthur Beer
Scagell, Robin, *Cambridge Guide to Stargazing with Your Telescope* (Cambridge University Press, 2000)
Schott Glaswerke, *Optical Glass* (Schott Optics Division, Mainz, 1992 [Cat.])
Sidgwick, John Benson, *Observational Astronomy for Amateurs* (Enslow, U.S. & U.K., 1982, 4th ed.)
Suiter, Harold Richard, *Star Testing Astronomical Telescopes* (Willmann-Bell, Richmond, 1994)
Taylor, H Dennis, *The Adjustment and Testing of Telescope Objectives* (Sir Howard Grubb, Parsons & Co – Andrew Reid, Newcastle upon Tyne, 1946)
Texereau, Jean, *How to Make a Telescope* (Willmann-Bell, Richmond, 1984, 2nd ed) trans. Allen Strickler
Todd, David, *A New Astronomy* (American Book Company, New York, 1897)
Twyman, Frank, *Prism and Lens Making* (London, Adam Hilger/IOP, 1989)
United States Navy, Bureau of Personnel, *Basic Optics and Optical Instruments* (Dover, New York, 1969 – reprint of Navy Training Course NAVPERS 10205, 1966)
Wallace, Brad D, and Provin, Robert B, *A Manual of Advanced Celestial Photography* (Cambridge University Press, 1988)
Warner, Deborah Jean and Ariail, Robert, *Alvan Clark & Sons, Artists in Optics* (Willmann-Bell, Richmond,1995, 2nd ed)
Webb Society, *Webb Society Deep-Sky Observer's Handbook*, Kenneth Glyn Jones, ed. (Enslow Publishers, U.K & U.S, 1986) Vol. I, 2nd ed.
Wood, FB, Ph.D. and Size, JW, *The Brashear 18-inch Refractor Telescope* (University of Pennsylvania, Flowers Observatory, 1963)
Worley, Charles E, *Visual Observing of Double Stars* (Sky Publishing, Cambridge, MA, ed. 1979 [article reprint])

Index

Notes: Page numbers in *italics* are illustrations.